LINEAR INTEGRATED CIRCUIT APPLICATIONS

G.B. Clayton

Department of Physics
Liverpool Polytechnic

M

© G. B. Clayton 1975

All rights reserved. No part of this publication may be reproduced or transmitted, in any form or by any means, without permission.

This book is sold subject to the standard conditions of the Net Book Agreement

First published in 1975 by

THE MACMILLAN PRESS LTD

London and Basingstoke
Associated companies in New York Dublin
Melbourne Johannesburg and Madras

SBN 333 18290 1 Case
SBN 333 18722 9 Limp

The paperback edition of this book is sold subject to the condition that it shall not, by way of trade or otherwise, be lent, resold, hired out, or otherwise circulated without prior consent in any form of binding or cover other than that in which it is published and without a similar condition including this condition being imposed on the subsequent purchaser.

Typesetting by Martin Dawson Coldtypesetter, Aberdeen and printed in Great Britain by
Butler and Tanner Ltd., Frome

LINEAR INTEGRATED CIRCUIT APPLICATIONS

Preface

The manufacture of circuits in integrated form dates from about 1959; the technology had its first impact in the field of digital electronics and it is here that the most impressive developments have taken place. It is only comparatively recently, after the general acceptance of the first linear i.c.'s, the operational amplifiers, that manufacturers have turned their efforts to the development of a variety of linear integrated circuit devices. Attention has naturally been directed to the linear devices which lend themselves to an economic utilisation of the i.c. process because of their potentially large volume production requirements. Both linear and digital i.c.'s simplify the design of electronic systems, they free the engineer from much detailed design work and allow him to concentrate on the problems of the total system.

In a rapidly expanding subject like electronic measurements there is always a difficulty in keeping in touch with the newer devices and techniques that are available. This book is an attempt to remedy the situation, it highlights those linear integrated circuit devices that are likely to be of most use in signal measurement and processing systems and shows how they can be used to perform the functional operations.

A practical approach is emphasised throughout the book. This is intended to encourage the reader to try out the devices for himself. Linear integrated circuit manufacturers strive to make their products with the greatest possible user convenience. The era of 'plug it in and it plays' has almost arrived. It is now possible, using linear i.c.'s, to build electronic systems to a degree of complexity and precision of operation not formerly possible with discrete component systems. Even in systems formerly implemented using discrete components the use of i.c.'s and modules will invariably be found to provide overall cost/performance advantages. There are savings in component costs when compared with discrete component designs of equivalent performance, there are also cost savings in assembly, inventory, incoming inspection, maintenance, etc. Performance advantages do not rest only in electrical performance but include mechanical performance, environmental performance (the effect of temperature, humidity etc.) and reliability performance.

The most widely used linear integrated circuit is the operational amplifier;

the operational amplifier is likely to remain in the forefront of linear system design for some considerable time, until perhaps the individual i.c.'s in measurement systems are absorbed in large scale integrations of the complete system. The applications of operational amplifiers are many and varied; they have been treated in two other books by the author[1,2]. In this book the applications of operational amplifiers as measurement amplifiers and the use of operational amplifiers in active filter circuits are dealt with. The remainder of the book is concerned with the more recently introduced linear integrated circuits, monolithic integrated circuit modulators, four quadrant multipliers, timers, waveform generators and phase locked loops. The book describes the principles underlying the operation of the devices and their applications in performing some of the functions required in signal measurement and processing systems.

Numerical exercises are included at the end of each chapter of the book. The exercises are designed to test comprehension and to give familiarity with design equations and with the order of magnitude of the components that are used in practical device applications. The book should prove useful to both the practising experimental scientist and the undergraduate student in a scientific or engineering discipline. The practical approach which it adopts should serve as a balance to the rather intensive theoretical treatment given in many undergraduate courses in electronic engineering.

Acknowledgements are made to the following manufacturers for information provided in their technical literature: Analog Devices, Motorola, Signetics, Burr Brown, Fairchild Semiconductor, R.C.A., Exar, Silicon General.

GBC

References

1. G.B. Clayton *Operational Amplifiers* Butterworth London (1971)

2. G.B. Clayton *Experiments with Operational Amplifiers* Macmillan (1975)

Contents

1 Measurement Circuits   1

1.1	Differential Amplifiers for Measurement Applications	2
1.1.1	One Amplifier Differential Circuit	2
1.1.2	Further Differential Circuits	4
1.1.3	An Integrated Circuit Instrumentation Amplifier	9
1.2	Modifying the Output Characteristics of an Instrumentation Amplifier	14
1.2.1	Use of Source and reference terminals	14
1.3	When is a Differential Measurement Amplifier Required?	17
1.4	Floating Network Measurements	23
1.5	Bridge Read-out Amplifiers	24
1.6	Photo-cell Amplifiers	30
1.6.1	Photo-Voltaic Cell Amplifiers	30
1.6.2	Photo-Diode Amplifier	31
1.6.3	Photo-Conductive Cell Amplifier	31
1.7	Charge Amplifiers	32
1.7.1	Capacitive Transducer Amplifier	34
1.8	Voltage and Current Meter Circuits	39
1.8.1	D.C. Voltage Measurements	40
1.8.2	D.C. Current Measurement	41
1.8.3	A.C. Measurements	44
Exercises 1		46

2 Some Signal Processing Applications   52

2.1	Active Filter Using Operational Amplifiers	52
2.2	Low Pass Active Filters	53
2.3	High Pass Active Filters	59
2.4	Band Pass Active Filters	62
2.4.1	VCVS Band Pass Filter	62
2.4.2	A Negative Inmittance Converter Band Pass Filter	65
2.5	Filter Realisations Using Analogue Computer Techniques	69

2.6		Band Reject Filters	73
	2.6.1	Twin 'T' Band Rejections Filters	76
	2.6.2	Practical Considerations Governing the Choice of Q	78
2.7		Multipole Filters – Different Types of Response	81
	2.7.1	Some Filter Designs	82
	Exercises 2		89

3 Monolithic Timing and Waveform Generator Devices 91

3.1		Monolithic Timing Circuits	91
	3.1.1	The 555 Timer, Free running Operation	91
	3.1.2	The 555 Timer, Monostable Operation	96
	3.1.3	Sequential Timing	99
	3.1.4	Keyed Oscillator	100
	3.1.5	Fixed Frequency Variable Duty Cycle Oscillator	101
3.2		Monolithic Waveform Generators	101
	3.2.1	The 566 Waveform Generator	101
	3.2.2	Single Cycle and Gated Operation of the 566	105
	3.2.3	The 566 Used as a Ramp Generator	107
	3.2.4	Asymmetrical Waveforms with the 566	107
	3.2.5	F.M. Generation with two 566's	109
	3.2.6	The Intersill 8038 Waveform Generator	110
	3.2.7	Frequency Modulation and Frequency Sweeping of the 8038	116
	Exercises 3		117

4 Variable Transconductance Devices 119

4.1		The Variable Transconductance of a Bipolar Transistor	119
4.2		Using Variable Transconductance for Gain Control	120
4.3		Controlled Gain I.C. Devices	125
	4.3.1	Controlled Operational Amplifiers	126
	4.3.2	The Gate Controlled, Two Channel, Wide Band Amplifier type MC 1545	128
4.4		Balanced Modulators	128
	4.4.1	The Balanced Modulator type SG1402	131
	4.4.2	The Balanced Modulator type MC1596	133
4.5		Variable Transconductance Linear Multipliers	136
	Exercises 4		142

5 Variable Gain Devices – Practical Considerations   143

5.1	Controlled Operational Amplifiers	143
5.2	The Two Channel, Gate controlled Wide Band Amplifier, type 1545	148
5.2.1	Parameter Measurements for the 1545 Device	148
5.2.2	Applications of the 1545 Device	151
5.3	Modulator Applications	155
5.3.1	Modulation Processes	155
5.3.2	Modulator Circuits	157
Exercises 5		164

6 Four Quadrant Linear Multipliers – Practical Considerations and Applications 165

6.1	Practical Multipliers – Departures from Ideal Behaviour	165
6.1.1	Feedthrough and Offsets	168
6.1.2	Scale Factor and Scale Factor Errors	169
6.1.3	Multiplier Non-Linearity	171
6.1.4	Total D.C. Error	173
6.2	Multiplier Test Circuits	173
6.2.1	Basic Circuit Arrangements	176
6.2.2	Measurement of Input Offsets and Bias Currents	177
6.2.3	Measurement of Non-Linearity	178
6.2.4	Frequency Response Characteristics	179
6.3	Multiplier Applications	180
6.3.1	Squaring, Dividing, Square Rooting	180
6.3.2	Mean Square and Root Mean Square	185
6.3.3	Power Measurement	188
6.3.4	Automatic Level Control Applications	191
6.3.5	Voltage Controlled Quadrature Oscillator	194
6.3.6	Further Computation Circuits	194
6.3.7	Modulator/Demodulator Applications	202
Exercises 6		207

7 Phase Locked Loops   208

7.1	Phase Locked Loop Building Blocks	208
7.2	The Phase Lock Loop Principle	209
7.3	Measurement of Lock and Capture Range – Display of Capture Transient	214

7.4		Parameters Determining Lock and Capture Range	221
	7.4.1	Phase Detector Conversion Gain K_d	221
	7.4.2	Low Pass Filter and Amplifier	222
	7.4.3	VCO Conversion Gain K_0	222
7.5		Dynamic Behaviour of the Locked Loop	224
	7.5.1	First Order Loop	227
	7.5.2	Second Order Loop	228
	7.5.3	Measurement of Dynamic Response	233
7.6		Modifying the Loop Characteristics	236
	7.6.1	Reducing the Lock Range of the 565	236
	7.6.2	Increasing the Lock Range of the 565	238
7.7		Phase Locked Loop Applications	239
	7.7.1	F.M. Demodulation	240
	7.7.2	A.M. Demodulation	241
	7.7.3	Phase Modulation	242
	7.7.4	Frequency Synthesis	242
	7.7.5	Frequency Translation	244
Exercises 7			246

Appendix   248

Answers to Exercises   261

Index   265

1. Measurement Circuits

Electronic measurement techniques play a major role in all branches of science in both research and industrial environments. In measurement systems the primary electrical signal source is generally of very low power and one of the most important elements in the system is the amplifier which performs the initial conditioning of the input signal. The accuracy of the complete system is very much dependent upon this first processing operation and the choice of a suitable amplifier configuration to perform the task is very important.

Amplifiers used to perform initial signal processing are sometimes referred to as 'Measurement Amplifiers', 'Data Amplifiers' or 'Instrumentation Amplifiers'. Operational Amplifiers can be configured into Instrumentation Amplifiers by the use of appropriate negative feedback circuits and some measurement applications can be perfectly adequately performed by a single operational amplifier. Difficulties arise when input signals are produced by transducers which, because of the constraints of the system, must be located a considerable distance from the amplifier. In this type of situation induced pick-up in long leads and unavoidable ground loops can give rise to signals which are many times greater than the transducer output signal which it is required to measure. In an attempt to minimise unwanted signals, measurement techniques based upon differential amplifiers are used, the underlying idea being that if the unwanted signals are transformed into common mode signals they can be rejected by the differential amplifier.

The design of an accurate measurement circuit requires a systematic appraisal of the complete system consisting of primary stimulus (the physical quantity to be measured), the transducer (used to convert the primary stimulus into an electrical signal) and the amplifier (used to step up the power level of the transducer output). The measurement circuit designer must have a sound knowledge of the nature and limitations of available transducers if he is to choose one which is suitable for use in the particular environment in which the measurements must be made. There are many different types of transducer in common use; thermocouples, resistive transducers, photoelectric transducers, piezoelectric transducers.

Some systems do not require a transducer as such for the physical variable under investigation is already in the form of an electrical signal. Examples are to be found in biological and medical instrumentation where electrical signals indicative of physiological processes are produced by living tissue. In this type of measurement, electrical contacts in the form of electrode probes are used and the signal source is essentially the living tissue.

A knowledge of the electrical output characteristics of the signal source (the transducer), is essential in the choice of a suitable measurement amplifier. Is the source a voltage or current source?; what is its output impedance?; what is the expected amplitude range and frequency content of the signal produced by the source? An answer to all these questions must be found if the choice of a suitable measurement amplifier is to be made. Many measurement circuits, as previously mentioned, are based upon differential input amplifiers and it is therefore appropriate to consider the design of such amplifiers in some detail before considering examples of specific measurement circuits.

1.1 Differential Amplifiers for Measurement Applications

The performance requirements of a differential input amplifier for use in measurement applications are generally focussed upon providing high input impedance and high Common Mode Rejection Ratios. A single operational amplifier can be connected as a differential input amplifier but, apart from the simplicity of the circuit, the arrangement, as will be shown, has several drawbacks which make it unsuitable for accurate signal conditioning. It is however, possible to improve on the single amplifier circuit in various ways by developing a differential configuration around two separate operational amplifiers instead of one and some of these possible circuits will be discussed. An alternative solution, perhaps the most economic one if design time and effort is taken into account, is to use a purpose-built instrumentation amplifier. Such closed loop gain blocks with differential inputs, high input impedance excellent CMRR and accurate gain have been available for some time in modular form, they are now available as single chip integrated circuits, for example Analogue Devices type AD520. The characteristics of the AD520 instrumentation amplifier will be discussed.

1.1.1 *One Amplifier Differential Circuit*

The limitations of the one amplifier differential circuit illustrated in figure 1.1 will be discussed. Assuming ideal amplifier performance and exact resistor proportioning the circuit performance is governed by the equation,

$e_0 = (e_2 - e_1) \dfrac{R_2}{R_1}$ and the ideal circuit completely rejects common mode

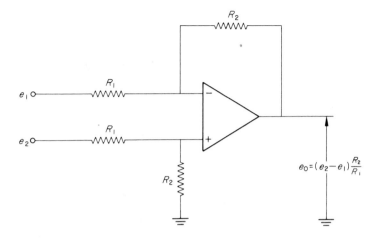

Fig. 1.1   One-amplifier differential circuit

input signals. The differential input resistance of the circuit is $2R_1$ and its common mode input resistance is

$$\tfrac{1}{2}\left[R_1 + \frac{R_2\,R_{cm}}{R_2 + R_{cm}}\right]$$

where R_{cm} is the common mode input resistance of the amplifier itself.

In a practical circuit difficulties arise if, in an attempt to step up the input resistance of the circuit, large values are used for resistors R_1, since large values of R_1 result in increased offset and drift. The very small bias current requirements of F.E.T. input amplifiers allows their use with large values for resistors but only at the expense of a possible increase in noise and a decrease in bandwidth due to stray capacitors.

The CMRR of the practical circuit is critically dependent upon resistor tolerance. If resistors are not perfectly matched a common mode input signal gives rise to a differential signal at the amplifier input terminals and this appears amplified at the output. Because of this effect, the common mode rejection ratio of the circuit can be considerably less than the common mode rejection ratio of the amplifier used in the circuit. It is, of course, theoretically possible to trim resistors so that the differential signal injected by a common mode signal, because of resistor mismatch, is of opposite polarity to the common mode error voltage due to amplifier non-infinite CMRR. In this way a theoretically infinite CMRR for the circuit can be obtained. In practice, this method can tweak-up the overall CMRR of the circuit by an order of magnitude greater than the CMRR of the amplifier used in the circuit. However, there are several hazards to the technique. For example, if the external resistors drift away from their 'tweaked up' values, either through temperature instability, or ageing, then the amplifier's common mode compensation drifts correspondingly. A compensating technique depends upon the constancy of the amplifier's internal CMRR, but in fact, this parameter varies in response to several factors. The CMRR of an amplifier is dependent upon the amplitude of the applied common mode signal, it also varies with output loading and common mode frequency.

Difficulties arise if a variable gain differential amplifier is required, if resistors R_2 in figure 1.1 were to be replaced by a ganged pair of variable resistors it would be virtually impossible to maintain a high CMRR because of the difficulty of maintaining accurate tracking of the two variable resistors. A method of adjusting the gain of the one amplifier differential circuit without altering its CMRR and using a single variable resistor is shown in figure 1.2. The circuit has the disadvantage of requiring six precision resistors and the gain varies non-linearly with the variable resistor setting.

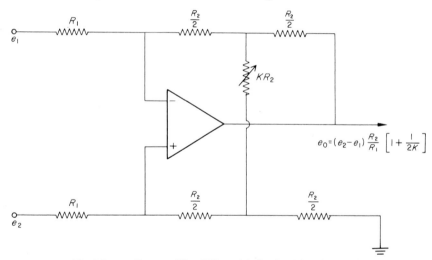

Fig. 1.2   One-amplifier differential circuit with variable gain

1.1.2 Further Differential Circuits

It is possible to improve on the single amplifier circuit in various ways by developing a differential configuration around two separate operational amplifiers instead of one. Differential amplifier circuits based upon non-inverting amplifiers may be expected to benefit from the high input impedance of such circuits, which is obtained without the use of high value resistors and consequent bandwidth limitations. Differential circuits based upon two inverting amplifier circuits can be expected to handle large common mode input voltages and to afford immunity to the operational amplifiers internal common mode errors.

A two-amplifier differential circuit based upon non-inverting amplifiers is shown in figure 1.3; it may be regarded as two coupled followers with gain. If e_2 is zero the second amplifier acts as a simple inverting amplifier with gain

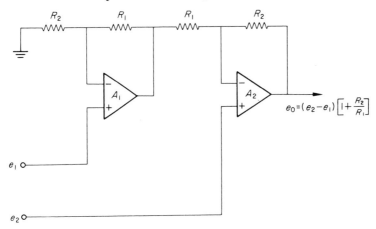

Fig. 1.3   High input impedance differential amplifier

minus $\dfrac{R_2}{R_1}$ operating upon the output of the first amplifier which is a follower of gain $(1 + R_1/R_2)$. The complete gain operating upon the input signal e_1 is thus

$$\left[1 + \frac{R_1}{R_2}\right] \times - \frac{R_2}{R} = -\left[1 + \frac{R_2}{R_1}\right]$$

If the input signal e_1 is zero the output of amplifier A_1 is zero and A_2 acts as a follower with gain $1 + R_2/R_1$ acting upon the input signal e_2. The output from the circuit may be written as

$$e_0 = (e_2 - e_1)\left[1 + \frac{R_2}{R_1}\right]$$

If $e_2 = e_1$ the output is theoretically zero but in practice the CMRR of the circuit is critically dependent upon resistor matching.

The circuit possesses the high input impedance characteristic of the followers but it is susceptible to the internal common mode errors of the amplifiers and the input common mode range is limited by the amplifiers input common mode range. The circuit can be given a variable gain facility without affecting its CMRR by connecting a resistor as shown in the circuit of figure 1.4. In this circuit the output voltage is determined by the relationship

$$e_0 = (e_2 - e_1)\left[1 + \frac{R_2}{R_1} + 2\frac{R_2}{R_3}\right]$$

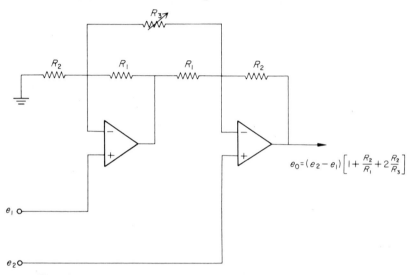

Fig. 1.4   High input impedance differential amplifier with variable gain

Note that the gain varies non-linearly with the magnitude of the gain control resistor R_3.

A two-amplifier differential circuit based upon the inverting configuration is shown in figure 1.5. The output signal of the second amplifier

$$e_{02} = -\left[\frac{R_2}{R_1} e_1 + n \frac{R_2}{R_1} e_{01}\right]$$

But $e_{01} = -\dfrac{e_2}{n}$

Thus

$$e_{02} = (e_2 - e_1)\frac{R_2}{R_1}$$

Linear Integrated Circuit Applications 7

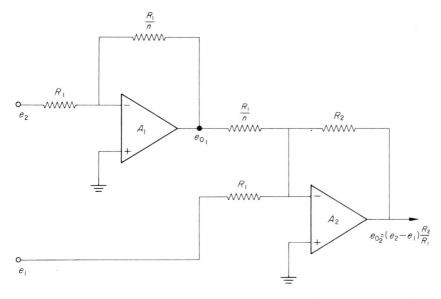

Fig. 1.5   Differential amplifier with large common mode range

By suitable proportioning of resistor values the circuit can be given quite a large common mode range. The common mode range is approximately $(\pm) n$ times the output voltage capability of the amplifiers. The CMRR of the circuit is critically dependent upon resistor tolerances but it does not depend upon the internal CMRR of the amplifiers. The differential input resistance of the circuit and the common mode input resistance of the circuit are both dependent upon the value of the resistors R_1. Bias current and bandwidth limitations due to stray capacitance place a practicable limit upon the magnitude of the input resistors R_1.

Another two-amplifier differential circuit based upon non-inverting amplifiers is shown in figure 1.6. The arrangement has the advantage of eliminating the dependence of CMRR on resistor matching, and providing a differential output signal which is ideal for driving isolated loads. A brief analysis of the action of the circuit is instructive in that it provides an insight into the circuit operation.

If the operational amplifiers are assumed to behave ideally the same current I_1 must flow through resistors R_1, R_2, R_3,

Thus

$$I_1 = \frac{e_{01} - e_1}{R_1} = \frac{e_1 - e_2}{R_2} = \frac{e_2 - e_{02}}{R_3}$$

which gives

$$e_{01} = e_1 \left[1 + \frac{R_1}{R_2}\right] - e_2 \frac{R_1}{R_2}$$

$$e_{02} = e_2 \left[1 + \frac{R_3}{R_2}\right] - e_1 \frac{R_3}{R_2}$$

Thus

$$e_{01} - e_{02} = (e_1 - e_2)\left[1 + \frac{R_1}{R_2} + \frac{R_3}{R_2}\right]$$

and the differential gain is

$$A_{\text{diff}} = \left[1 + \frac{R_1}{R_2} + \frac{R_3}{R_2}\right]$$

The gain can be adjusted by means of the single variable resistor R_2.

Note if $e_1 = e_2 = e_{cm}$
then $e_{01} = e_{cm}$ and $e_{02} = e_{cm}$

Common mode signals are passed at unity gain by both amplifiers and a common mode input signal does not give rise to a differential output signal. If the circuit is used to drive an isolated load it provides a theoretically infinite CMRR which is quite independent of resistor values. In practice, CMRR when driving an isolated load will be very large, but not infinite, because of possible mismatch in the internal CMRR of the amplifiers. A common mode signal e_{cm} may be expected to give rise to a differential output signal of value.

$$e_{cm}\left[\frac{1}{\text{CMRR}_{A_1}} - \frac{1}{\text{CMRR}_{A_2}}\right]$$

The differential output given by the circuit of figure 1.6 can be converted into a single ended output by following the cross-coupled non-inverters with a one amplifier differential circuit, the arrangement of which is shown in figure 1.7. The cross-coupled followers provide a differential gain but pass common mode signals at unity gain; the overall CMRR of the circuit is thus effectively the CMRR of the one-amplifier differential circuit multiplied by the differential gain of the cross-coupled follower circuit which precedes it. This suggests that the circuit's CMRR can be maximised by assigning most of the overall gain to the cross-coupled follower stage. In fact, this cannot be pushed too far without running into difficulty. A high first-stage gain will either restrict the level of common mode voltage or cramp the circuits' dynamic swing, or both. The followers pass common mode signals at unity gain, thus signal variations at e_{01} and e_{02} take place about the input common mode level and a restriction is set by the output voltage rating of the amplifiers.

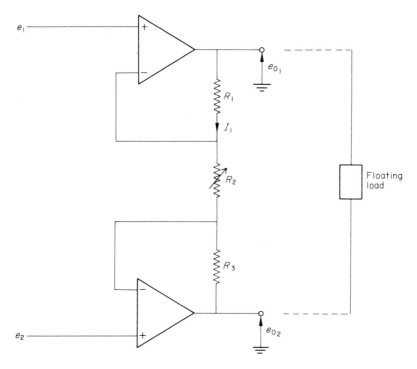

Fig. 1.6   Differential input, differential output, high input impedance amplifier

The circuit configuration of figure 1.7 is capable of excellent performance as a differential input instrumentation amplifier, detailed performance characteristics and performance errors are of course dependent upon the performance characteristics of the operational amplifiers used in the circuit. The drift performance of the circuit is determined by the drift characteristics of the followers, a matching of the drift performance of these amplifiers would have the effect of cancelling out drift. Identical drift in the followers represents a common mode signal and as such is rejected by the subtractor circuit.

1.1.3 *An Integrated Circuit Instrumentation Amplifier*

As an alternative to the user-connected differential amplifier circuits discussed in the previous section, the circuit designer may find a solution to his measurement amplifier requirements by using a purpose-built instrumentation amplifier. There is now a wide variety of data amplifiers on the market, the more sophisticated ones providing facilities such as digitally programmed gain or autoranging, but these 'extras' as is to be expected, tend to increase the price.

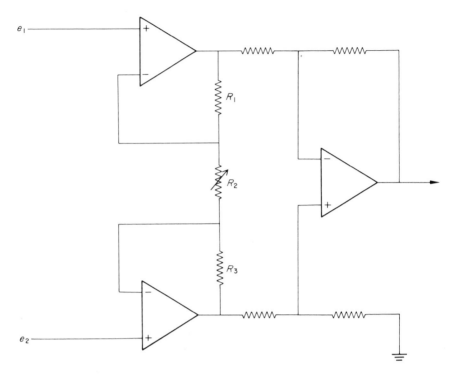

Fig. 1.7   Differential input, single-ended output, high input impedance amplifier

However, simple basic instrumentation amplifiers with very good performance characteristics are now available at a low cost and this must encourage their use in many measurement applications. An example of such a device is the *Analog Devices* Type AD 520, this is a differential instrumentation amplifier on a single silicon chip; its action and characteristics will be discussed.

Confusion in terminology between operational amplifiers and instrumentation amplifiers can often arise. Operational amplifiers are basically open loop gain blocks, they may be single ended or differential and they are used to drive external feedback networks which define circuit performance. By contrast, instrumentation amplifiers are closed loop gain blocks, they have a differential input, high input impedance, high CMRR and accurately specified gain.

The AD 520 instrumentation amplifier has a differential input impedance and common mode input impedance of 2×10^9 Ω and CMRR of typically 80 dB at gain unity and 110 dB at gain 1000. Its gain is adjustable in the range 1 to 1000 by choice of a single external gain setting resistor in the range 100Ω

to 100kΩ and the gain linearity of the device is good with non-linearity expressed as a percentage of full scale typically 0.02 per cent. It has small input bias and offset currents, (30nA 10nA), and at gain 500, has both a small signal and full power frequency response up to 60kHz. In addition it provides performance features not readily obtainable with the user-connected instrumentation amplifiers. An output reference terminal allows the output to be biased independently of the gain setting, a useful feature for positioning chart recorder writing elements. A high impedance sense terminal allows the circuit's feedback to be derived from either the output terminal or an arbitrary external point. The latter mode allows the AD 520 to be used as a 'constant' current generator, or with an external inside the loop booster for increased output power.

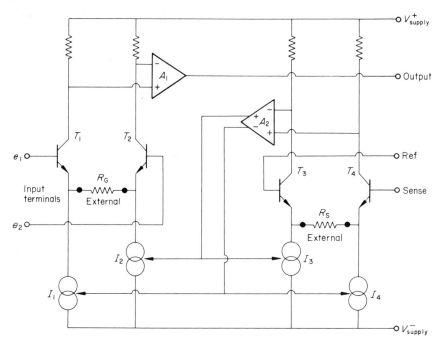

Fig. 1.8 Simplified functional schematic of AD 520 instrumentation amplifier

The action of the AD 520 may be understood in terms of the simplified functional schematic given in figure 1.8. Readers interested in the complete circuit schematic are referred to the amplifier data sheet. The device contains two differential amplifiers, the main signal amplifier A_1 and a second differential amplifier A_2 which is used to control four current sources, I_1, I_2, I_3 and I_4. I_2 and I_1 track I_3 and I_4. Unlike operational amplifier feedback circuits, in which negative feedback is applied by the external connection of components between output and input terminals, the feedback loop in the AD 520 device is internal. In normal applications the feedback loop is completed by connecting the output terminal of the amplifier to the so-called 'sense' terminal. The high gain differential input amplifiers A_1 and A_2 are then both in enclosed feedback loops. The action of the feedback is to force the differential input signals applied to the amplifiers towards zero, and this point should be kept clearly in mind when attempting to visualise the action of the complete circuit.

Consider a difference signal $e_1 - e_2$ applied to transistors T_1 and T_2. If I_1 and I_2 remained constant the collector currents of T_1 and T_2 would be unbalanced by an amount $(e_1 - e_2)/R_G$ which in turn would apply a differential input signal to amplifier A_1. The output signal of A_1 acts via the 'sense' terminal and the control amplifier A_2 works to equalise the collector currents of T_1 and T_2. It does this by adjusting the current sources I_1 and I_2. At the same time the output signal of A_1 applied to the 'sense' terminal tries to unbalance the collector currents in T_4 and T_3, (by an amount $(V_{sense} - V_{ref})/R_S$) but A_2 works towards maintaining its zero differential input signal by adjusting the current sources I_4 and I_3.

Assuming A_1 and A_2 have very high gain we may neglect their differential input signals as insignificant errors and write the equilibrium conditions in the circuit as

$$I_{c1} = I_{c2}$$

or $$I_1 + \frac{e_1 - e_2}{R_G} = I_2 - \left[\frac{e_1 - e_2}{R_G}\right]$$

and $$I_2 - I_1 = 2\left[\frac{e_1 - e_2}{R_G}\right]$$

Collector currents of T_3 and T_4 are equalised

$$I_{c4} = I_{c3}$$

or $$I_4 + \left[\frac{V_{sense} - V_{ref}}{R_S}\right] = I_3 - \left[\frac{V_{sense} - V_{ref}}{R_S}\right]$$

and $$I_3 - I_4 = 2\left[\frac{V_{sense} - V_{ref}}{R_S}\right]$$

Current sources track one another, $I_2 = I_3$ and $I_1 = I_4$

Thus $$\frac{V_{sense} - V_{ref}}{e_1 - e_2} = \frac{R_S}{R_G} = G$$

the closed loop gain of the complete system. With the sense terminal connected to the output terminal $V_{sense} = e_0$ and

$$e_0 = V_{ref} + G(e_1 - e_2)$$

Note that the gain is set by two external resistors R_G and R_s and an output offset which is independent of the gain setting may be obtained by returning the reference input terminal to any required voltage within the output capability of the amplifier ($\pm$ 10 volts with $V_s \pm$ 15 volts)

Setting-up Procedure. Typical external connections required for the AD 520 are shown in figure 1.9, the following adjustment procedure is suggested.

Nulling of the output offset is accomplished as follows; connect a 50 ppm 1kΩ trimming potentiometer between pin 2 and pin 3. With the input terminals

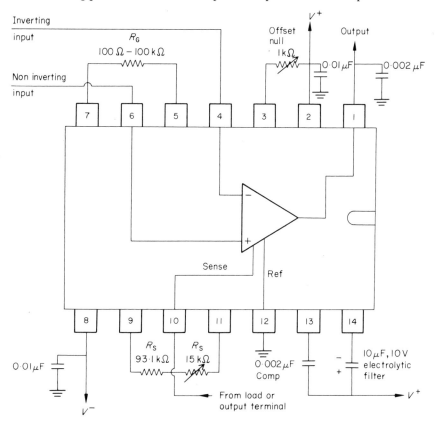

Fig. 1.9   Typical operating connections for AD 520 instrumentation amplifier

pin 4 and pin 6 both grounded, adjust the trimming potentiometer for zero output. This procedure should be repeated if gains are changed since the offset, although minimised by close resistor and transistor matching within the amplifier, will show slight variations with different gain connections. If output offsets are not critical a 400Ω fixed resistor must be substituted for the nulling procedure. Should the 1kΩ trimming potentiometer not yield sufficient resolution a fixed trimming resistor plus a smaller value potentiometer can be employed.

High accuracy gain calibration may be accomplished as follows; connect a precision (percentage as desired) 100kΩ resistor in the R_G position (between pins 5 and 7) and insert a fixed 93.1kΩ (standard value) 1 per cent metal film resistor in series with a 15kΩ trimming potentiometer in the R_S position (between pins 9 and 11). With pin 4 grounded drive pin 6 with a ± 10V d.c. signal and adjust the R_S potentiometer for a precise gain of unity measured by a zero indication of a voltmeter connected between pin 6 and pin 1 (the output). Once this gain adjustment is made R_S remains fixed and R_G can be selected for any desired gain within the 1–1000 range of the amplifier. The gain formula R_S/R_G is now precisely $G = 10^8/R_G$ within the tolerance of the gain calibration resistors used.

Frequency compensation of the AD 520 is provided by connecting a 0.002 μF capacitor between pin 13 and pin 2. This capacitor represents a lead in the overall response in the amplifier system, thus, if a slightly different frequency response is desired, the value of the capacitor can be raised or lowered to increase or decrease the bandwidth. A decrease in bandwidth can also be achieved by adding a resistor in series with the capacitor. While this is sometimes desirable to minimise high frequency noise in low speed applications, care must be exercised in the selection of the resistor since too high a value may cause frequency instability in the system. Increasing the bandwidth by raising the value of the compensating capacitor should be done carefully so as to minimise peaking. Frequency compensation with the standard 0.002 μF capacitor should be found suitable for the large majority of applications.

Note, that in addition to the frequency compensation capacitor it is always necessary to load the AD 520 with a 0.002 μF capacitor. In some applications, this minimum value will be provided by the load and/or a cable at the output of the amplifier making an additional 0.002 μF unnecessary.

1.2 Modifying the Output Characteristics of an Instrumentation Amplifier

1.2.1 *Use of Sense and Reference Terminals*

The provision of external sense and reference terminals in an instrumentation amplifier gives a facility for modifying the output characteristics of the amplifier. It allows an output offset control which is independent of gain, the inclusion of a current booster within the feedback loop to give increased output current and it permits the amplifier to be configured as a current feedback circuit for loads requiring a current drive.

In order to understand the action of sense and reference terminals it should be borne in mind that it is the voltage developed between these terminals that acts as the drive to the feedback network. The one-operational amplifier differential circuit of figure 1.1 and the output operational amplifier in the user connected instrumentation amplifier of figure 1.7 have the accessible sense and reference terminals shown in figure 1.10.

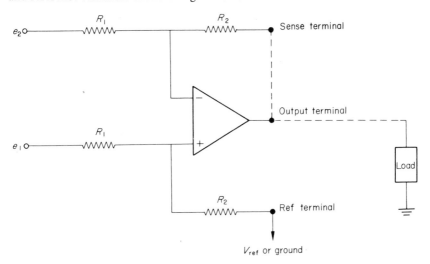

Fig. 1.10 Differential amplifier, sense and reference terminals

If the reference terminal is connected to a low impedance reference source V_{ref} and the sense terminal is connected to the output terminal of the amplifier, the output of the amplifier is determined by the expression

$$e_0 = \frac{R_2}{R_1}(e_1 - e_2) + V_{ref}$$

The circuit forces $V_{sense} - V_{ref}$ to be equal to the closed loop gain multiplied by the input difference voltage and in figure 1.10 $V_{sense} = e_0$. V_{ref} provides an output offset control but note that the internal impedance of the reference source causes a circuit unbalance which will degrade CMRR.

The operational amplifier differential circuit can be given an increased current output capability by including a current booster in the feedback loop; the sense terminal is connected to the output terminal of the current booster as shown in figure 1.11.

In current feedback configurations, used for loads requiring a current drive, the voltage $V_{sense} - V_{ref}$ is derived from a current sensing resistor connected in series with the load. Two arrangements are possible dependent upon whether the load is floating, (figure 1.12), or grounded, (figure 1.13). In both circuits $V_{sense} - V_{ref}$ is forced to be equal to the closed loop gain, R_2/R_1, multiplied by the input difference voltage.

Circuit performance equations are governed by

$$V_{sense} - V_{ref} = i_0 R = \frac{R_2}{R_1} (e_1 - e_2)$$

or

$$i_0 = \frac{1}{R} \frac{R_2}{R_1} (e_1 - e_2)$$

Care should be taken to ensure that $i_0 (R + Z_L)$ does not exceed the output voltage capability of the amplifier; watch out for inductive loads!

In the circuit of figure 1.12, the output current can be offset by returning the reference terminal to a low impedance reference voltage source instead of ground.

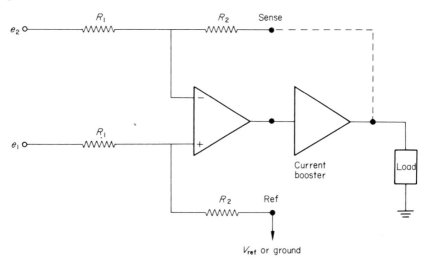

Fig. 1.11   Differential amplifier with increased current output

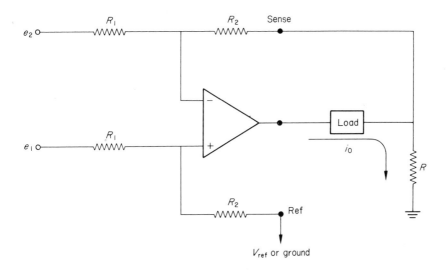

Fig. 1.12   Differential amplifier providing a current drive; load floating

The circuits of figures 1.12 and 1.13 have definite limitations which must be recognised; the performance equation is based upon the assumption of a negligible loading of the equivalent voltage sources driving V_{sense} and V_{ref}. This assumption may not be valid in a practical application. Any loading by the sense and reference terminals unbalances the circuit and degrades CMRR. The AD 520 instrumentation amplifier does not suffer from this limitation since its sense and reference terminals have a high input impedance ($5 \times 10^7 \Omega$), they are not directly associated with the differential input stage and therefore do not degrade CMRR. The AD 520 output can be modified in a manner similar to that shown in figures 1.10 to 1.13. The circuit arrangements are shown in figure 1.14, performance equations are readily derived by remembering that $V_{sense} - V_{ref}$ is forced to equality with $G(e_1 - e_2)$.

1.3 When is a Differential Input Measurement Amplifier Required?

The advantages obtained by using a measurement amplifier with a differential input are not always immediately obvious and indeed, some measurement applications can be handled quite satisfactorily with a single-ended input amplifier. It was mentioned earlier that differential input amplifiers are used to reject unwanted signals, the origin of such unwanted signals and the 'mechanics' of the rejection process are worthy of further consideration.

For the purpose of discussion consider a signal source which behaves as a

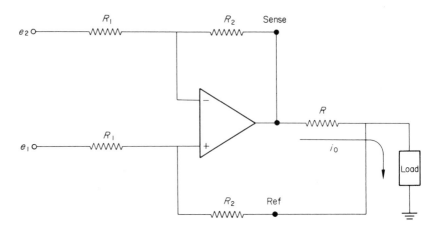

Fig. 1.13 Differential amplifier providing a current drive; load grounded

voltage source e_s of output resistance R_s. If the source is completely isolated from ground and is located at no great distance from the amplifier, a differential input amplifier will probably give no significant advantage over a single-ended one, provided certain elementary precautions are observed. Input leads should be as close together as possible, avoiding a large area input circuit loop which, cut by stray mains magnetic field, could result in the injection of unwanted mains hum. The signal source should not be separately grounded. Care should be taken to ensure that neither the power supply current nor the load current flow through the input signal ground connection. The proper connections are illustrated in figure 1.15.

Difficulties arise with a single-ended input amplifier when the signal source and remote amplifier are separately grounded. Even if the signal source is not physically connected to ground it is difficult to avoid some path to ground at the transducer site due to either leakage, or capacitance to ground, or both. The situation which exists when amplifier and source are separately grounded is illustrated in figure 1.16: there are now two potentially large area circuit loops, A B C D E F and A B E F. Any stray mains magnetic field which cuts these loops will induce a signal and cause a voltage to appear between the two ground points A and F. In addition to mains pick up there are many other possible contributions to a ground-to-ground potential difference when the ground points are located a considerable distance away from one another. Contributions include earth currents due to other electrical equipment, electrochemical effects, thermoelectric effects, created by temperature gradients in buried metals and soil elements and doubtless many

Linear Integrated Circuit Applications

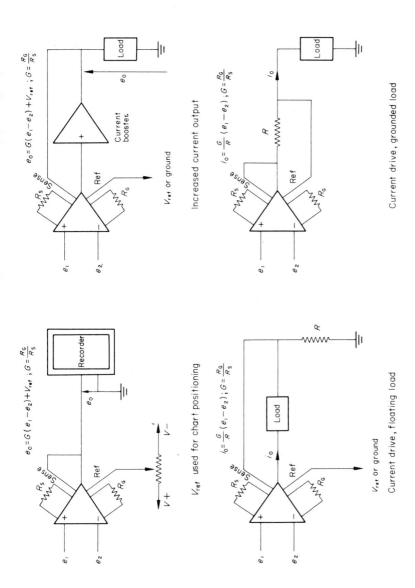

Fig. 1.14 AD 520 instrumentation amplifier; use of sense and reference terminals

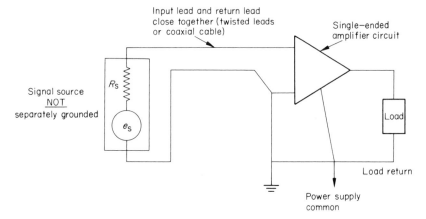

Fig. 1.15 Proper connections for single-ended measurement amplifier

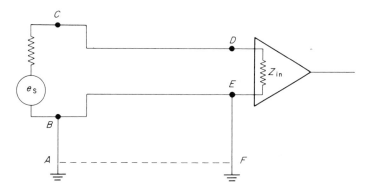

Fig. 1.16 Earth loops when source and single-ended amplifier are separately grounded

more possible contributions. In the equivalent circuit illustrated in figure 1.17 all contributions to ground potential difference are lumped together in a single source of e.m.f. e_g in series with Z_g the sum of all the impedances in the path B A F. Resistances R_L are included in the equivalent circuit to represent the resistance of the lines used to connect the signal source to the amplifier. In a practical situation, with long connecting lines, the lead resistances R_L are likely to be considerably greater than the ground-to-ground impedance. Consequently most of the ground voltage e_g appears between the points B and E and if Z_{in} is much bigger than $R_L + R_s$, as is normally the case, practically all of e_g is applied to the amplifier input terminals.

Consider now the arrangement in figure 1.18 where a differential input

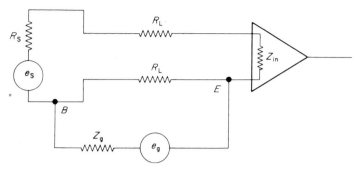

Fig. 1.17   Equivalent generator; used to represent unwanted ground to ground voltage

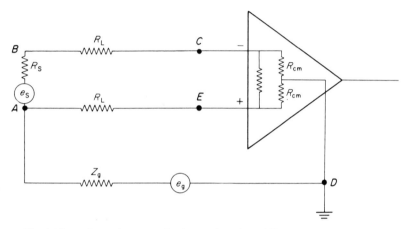

Fig. 1.18   Ground to ground voltage rejected by differential input amplifier

amplifier is used instead of the single-ended one. The ground-to-ground voltage now acts as a common mode input signal and as such is rejected because of the amplifiers large CMRR. Unfortunately, rejection is not complete even if the CMRR of the amplifier is infinite; any resistance unbalance in the paths A B C D and A E D causes part of the ground signal to be injected as a differential signal at the differential input terminals of the amplifier. In figure 1.18 the source resistance R_s represents an obvious source of unbalance but differences in lead resistance and in the common mode input resistances at the two input terminals represent other less obvious causes of unbalance. In figure 1.19 the net resistance unbalance is represented by R_u and the differential input voltage injected as a result of this unbalance is

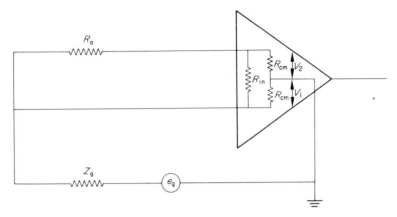

Fig. 1.19   Effect of resistance unbalance on CMRR

$$V_1 - V_2 = e_g \left[1 - \frac{R_{cm}}{R_{cm} + R_u} \right]$$

or

$$V_1 - V_2 \simeq e_g \frac{R_u}{R_{cm}} \quad \text{if } R_{cm} \gg R_u$$

This fraction of e_g is amplified by the differential gain of the amplifier in the same way as the desired input signal e_s. The common mode input signal e_g appears at the output of the amplifier multiplied by

$A_{diff} \dfrac{R_u}{R_{cm}}$ and the CMRR due to unbalanced source impedance is

$$\frac{A_{diff}}{A_{diff} \dfrac{R_u}{R_{cm}}} = \frac{R_{cm}}{R_u}$$

Theoretically it is possible to introduce a compensating resistance into the appropriate signal line in order to cancel out the effect of R_u, practically this is not always possible and in any case it is inconvenient. The solution is to use an amplifier with a very large value of common mode input resistance so that the CMRR due to source resistance unbalance, R_{cm}/R_u is as large as possible. Differential instrumentation amplifiers are typically rated for CMRR with a specified amount of source unbalance present, usually 1kΩ unbalance. In order to provide a CMRR of 100 dB with such an unbalance it requires a value of

$R_{cm} > \text{CMRR} \times R_u$, that is $R_{cm} > 10^8 \, \Omega$

A variety of measurement applications using both differential input instrumentation amplifiers and single operational amplifiers are given in the remainder of this chapter. In applications requiring a significant degree of accuracy a practical circuit implementation should always be preceded by a systematic appraisal of all error contributions. Identifying the most significant error terms may allow their reduction by the use of an alternative amplifier type or a different circuit configuration.

1.4 Floating Network Measurements

The high input impedance, both differential and common mode, and high CMRR of instrumentation amplifiers makes them suitable for measuring currents and voltages in unbalanced networks. The principle is illustrated in figure 1.20 where the measurement amplifier is connected across a current sensing resistor so that the output of the amplifier gives a measure of the branch current. The amplifier could equally well be used to measure the voltage across reactive elements. One of the high input impedance user-connected operational amplifier instrumentation amplifier configurations, or the AD 520 instrumentation amplifier can be used in this type of application.

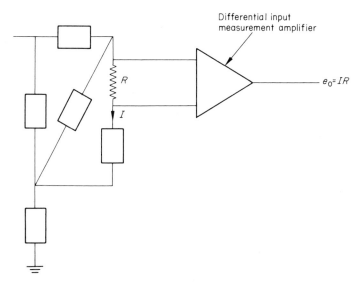

Fig. 1.20   Floating voltage measurement

1.5 Bridge Read Out Amplifiers

A wide range of transducers exist which consist essentially of a resistive element and resistive transducers are available which respond to temperature, light intensity, and physical strain. When precise measurements are to be made using resistive transducers, the transducers are normally included in the arms of a balanced bridge. Changes in the physical variable to which the transducer is sensitive cause an unbalance in the bridge, the extent of the unbalance being used to measure the change in the physical variable.

Consider, first of all, the basic passive bridge circuit. Either the out-of-balance voltage, figure 1.21a or the out-of-balance current, figure 1.21b, can be used as an indication of unbalance and hence as a measure of the fractional change a, in the resistance of the resistive transducer. But note that in both cases the bridge signal is not linearly related to a, an approximately linear relationship exists only for small values of a.

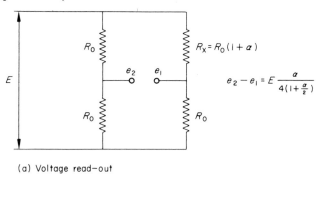

(a) Voltage read-out

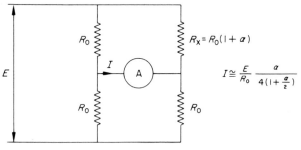

(b) Current read-out

Fig. 1.21 Passive bridges

There are a variety of ways of amplifying the out-of-balance signal given by a bridge, the choice of a suitable circuit configuration is very much dependent upon the detailed circumstances in the particular application. Some of the factors which determine the choice of circuit are: must the bridge voltage supply be grounded or can it float?; is the variable resistor grounded or can it float?; is a strictly linear conversion required? The extent of the physical separation between the bridge and the measuring amplifier, and hence a possible need for long connecting leads, is also an important consideration, there are other factors which will emerge from the discussion of bridge amplifier configurations which is now to be given.

A common application for differential input instrumentation amplifiers is for a straightforward amplification of bridge unbalance voltage, applications of this type are illustrated by the circuit shown in figure 1.22. In a typical application the bridge ratio resistors are a matched pair, while R_0 and R_x are nominally matched, generally varying according to the quantity to be measured (for example, elements of a strain gauge bridge). The output signal

$$e_0 = A_{\text{diff}} E \frac{\alpha}{4(1 + \frac{\alpha}{2})}$$

does not depend upon the bridge impedance level and so is not affected by changes in this impedance level, due to, for example, ambient temperature variations altering the resistance of all bridge elements. If bridge and amplifier are located a considerable distance apart the inevitable mains pick-up will be rejected by the high CMRR of the instrumentation amplifier.

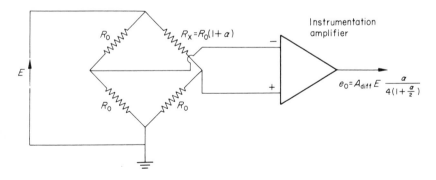

Fig. 1.22 Differential input instrumentation amplifier used for bridge read out

In applications which are less demanding of common mode rejection it is possible to use amplifiers that are less complex than a fully developed differential instrumentation amplifier. There are several circuit arrangements possible which use only a single operational amplifier.

A bridge circuit which gives a linear conversion for both small and large fractional changes in the variable resistor is shown in figure 1.23. The circuit is basically an application of the one-amplifier differential circuit, referring to figure 1.23 and assuming that the amplifier behaves ideally the following analysis holds

$$e_- = e_0 + \frac{(E - e_0) R_2}{R_1 + R_2}$$

$$e_+ = E \frac{R_4}{R_3 + R_4}$$

But $e_+ = e_-$

Substitution and rearrangement gives

$$e_0 = \left(\frac{R_4 - \frac{R_2}{R_1} R_3}{R_4 + R_3} \right) E$$

If R_3 and R_4 are made the bridge ratio resistors, $R_3 = R_4 = R$, (a matched pair), and the reference and variable resistors are connected in place of R_1 and R_2

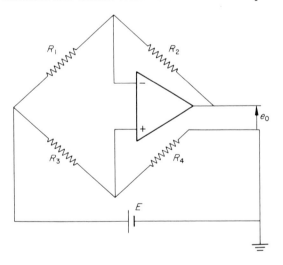

Fig. 1.23   Bridge amplifier linear read out

so that $R_1 = R_0$ and $R_2 = R_0(1 + \alpha)$ the circuit performance equation becomes

$$e_0 = \frac{\alpha}{2} E$$

Used in this way the circuit gives an output voltage which is linearly related to α, linearity is maintained for large deviations from bridge balance, but note that the unknown resistor must float.

An alternative way of using the circuit is to place the unknown resistor in the position occupied by R_4, making $R_3 = R_0$, $R_4 = R_0(1 + a)$ and $R_1 = R_2 = R$. Substitution of these values gives the relationship

$$e_0 = \frac{\alpha}{2 + \alpha} E$$

Note that the output now varies linearly only for small deviations in the unknown resistor. This method of connection is useful when the unknown resistor must be earthed, the amplifier output does not have to supply the current passing through the unknown resistor, thus large currents may be passed through it if this is required by the application. Both circuit arrangements give an output signal which is not dependent upon the bridge impedance level. However, the circuits do not provide amplification and the measurement of small resistance changes would probably require the use of an additional amplifier stage.

A circuit which makes use of a differential input operational amplifier to measure the out-of-balance current given by a bridge is shown in figure 1.24. The circuit forces the bridge unbalance voltage to zero and in doing so the

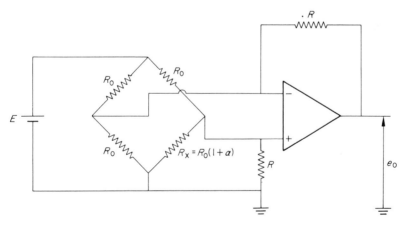

Fig. 1.24 Differential amplifier bridge current read out

amplifier generates an output signal which depends upon the bridge unbalance current. The circuit performance is governed by the equation

$$e_0 = E \frac{R}{R_0} \frac{\alpha}{1+\alpha} \frac{1}{\frac{(2+\alpha)}{1+\alpha} + \frac{R_0}{R}}$$

The performance is linear only for small deviations from balance. If $\alpha \ll 1$ and $R \gg R_0$ the performance equation reduces to the approximate form

$$e_0 = E \frac{R}{R_0} \frac{\alpha}{2}$$

An advantage of the arrangement is its comparative simplicity, it provides rejection of common mode signals to an extent dependent upon the CMRR of the amplifier. Common mode rejection is degraded by any source unbalance but the bridge behaves as an almost balanced source. The main disadvantage of the circuit is its sensitivity to the bridge impedance level. R_0 appears in the circuit performance equation so that gain is affected by any temperature-dependence of the bridge elements.

All the bridge circuits discussed thus far have used a differential input amplifier but in applications in which common mode rejection is not essential, because unwanted pick-up can be satisfactorily shielded, a single-ended circuit configuration becomes possible. A single-ended circuit allows the use of a single-ended chopper stabilised amplifier for very low drift errors. The arrangement shown in figure 1.25 is an example of an inverting bridge amplifier which can be implemented using a single-ended inverting operational amplifier. The circuit multiplies the bridge out of balance voltage by the closed loop gain $(1 + R_2/R_1)$ and the output voltage of the amplifier is governed by the relationship

$$e_0 = \left(1 + \frac{R_2}{R_1}\right) E \frac{\alpha}{4\left(1 + \frac{\alpha}{2}\right)}$$

Sensitivity is not dependent upon bridge impedance level but the circuit has the disadvantage of requiring a floating bridge supply, and like most of the other circuits, is linear only for small deviations from balance.

A simple circuit which can be used with a single-ended input amplifier is the so called 'half-bridge' arrangement illustrated in figure 1.26. Feedback holds the half-bridge output voltage at ground potential and the amplifier output voltage

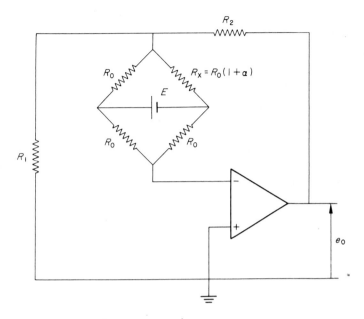

Fig. 1.25 Single-ended bridge amplifier

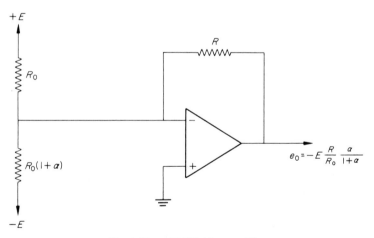

Fig. 1.26 Half-bridge amplifier

gives a measure of the short circuit output current of the bridge. The performance equation is

$$e_0 = -E \frac{R}{R_0} \frac{\alpha}{1+\alpha}$$

30 Linear Integrated Circuit Applications

Gain is dependent upon the bridge impedance level but note that the magnitude of the bridge supply voltage that can be used is not determined by common mode range limitation of the amplifier.

1.6 Photo-Cell Amplifiers

A variety of solid state light-sensitive transducers are in common use, in most applications the output given by these photo-sensitive devices requires amplification before it can be used to drive a recording instrument. Photo-generators display a voltage dependence of both speed and linearity and are best operated with a constant terminal voltage. They act essentially as current sources and the function of the measurement amplifier used with them is to perform a current-to-voltage conversion.

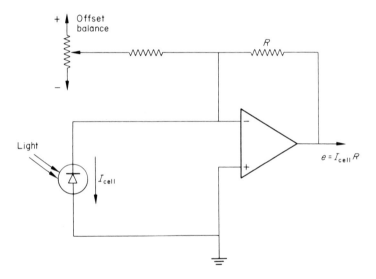

Fig. 1.27 Photovoltaic cell amplifier

1.6.1 *Photo-Voltaic Cell Amplifier*

The short circuit output current of a photo-voltaic cell is linearly related to the light intensity falling upon it, this means that for conversion linearity the cell should be operated with a zero terminal voltage. The condition can be realised by connecting the cell directly across the input terminals of an operational amplifier as shown in figure 1.27. Negative feedback forces the voltage across the cell to zero and causes the photo current to pass through the feedback resistor R; the amplifier develops an output voltage $e_0 = -I_{cell} R$. The value to be used for R is dependent upon cell sensitivity and should be chosen either for maximum dynamic range or for a desired scale factor. The choice of amplifier type is determined by allowable drift error as set by accuracy requirements. If a signal proportional to differential light intensity is required two cells can be used, connected as shown in figure 1.28.

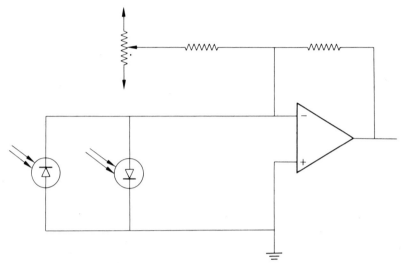

Fig. 1.28   Photovoltaic cell amplifier for measurement of differential light intensity

1.6.2 Photo-Diode Amplifier

A photo-diode is a p-n junction which is normally operated under reverse bias conditions when the reverse current is proportional to the light intensity falling upon the device. A typical photo-diode amplifier circuit is shown in figure 1.29. In addition to the desired light-sensitive current a photo-diode current contains leakage current components which can introduce significant errors at high temperatures. Some photo-diodes can be operated with zero terminal voltage when they act like a photo-voltaic cell and deliver a short circuit current which is unaffected by leakage currents and which is not significantly less than the output current with reverse bias. Such photo-diodes can be used in a circuit of the type shown in figure 1.27, reducing diode leakage currents by at least two orders of magnitude.

1.6.3 Photo-Conductive Cell Amplifier

Photo-conductive cells behave like resistive circuit elements with resistance

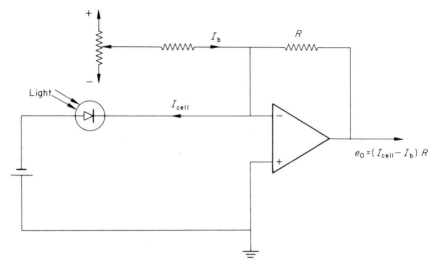

Fig. 1.29   Photo-diode amplifier

value varying according to the amount of illumination falling upon them. A simple photo-conductive cell amplifier is illustrated in figure 1.30a. Differential light intensity measurements can be made using two photo-conductive cells connected in a half bridge circuit as shown in figure 1.30b or they may be connected in one of the differential bridge circuits discussed in section 1.5 and used with a differential input amplifier in order to reject common mode pick up.

1.7 Charge Amplifiers

The ability to measure charge, that is, to integrate current, a task which can be performed by an operational integrator, is an important and useful measurement technique. Some transducers such as capacitive microphones, certain types of accelerometer, and piezoelectric pressure transducers, operate by producing a charge which is proportional to the measurement variable. Dosimetry is another field in which a charge measurement is required. The charge measured is proportional to the amount of radiation exposure received by a suitable material. Secondary radiation from the material (optical or atomic) is detected by a photo-multiplier or ion chamber whose current output is integrated to give charge which is proportional to the radiation exposure. In applications such as taking photo-micrographs with an electron microscope, current integration is used as a timing device. The electron beam is sensed with an electron multiplier

Linear Integrated Circuit Applications 33

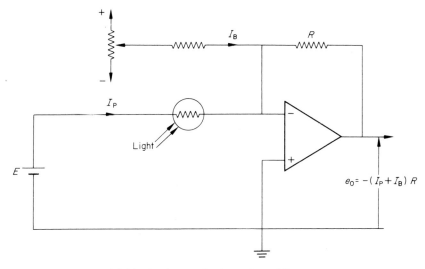

(a) Simple photoconductive cell amplifier

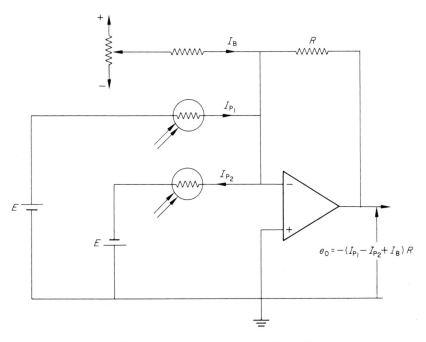

(b) Half-bridge photoconductive cell amplifier

Fig. 1.30   Photo-conductive cell amplifier

and fed into a current integrator with a level sensing device. Since the exposure time and integration time, are the same, the system can be used to time and control the electron microscope for correct exposure.

$$t = \frac{q}{i}$$

where q is the measurement charge, i is the electron beam current and t the exposure time period.

1.7.1 Capacitive Transducer Amplifiers

As in all measurement applications, processing of transducer signals requires an understanding of the electrical characteristics of the transducer in order that a measurement amplifier configuration which will accurately extract information from the transducer can be chosen. Piezoelectric transducers are essentially capacitive, they act by producing a charge which is proportional to the physical variable under investigation, for example acceleration in the case of a piezoelectric accelerometer. The electrical equivalent circuit of a piezoelectric transducer is illustrated in figure 1.31, the charge generated by the transducer is determined by

$$q = S_q\, m$$

where m is the measurement variable and S_q is the so-called charge sensitivity of the transducer. A charge q produces an open circuit voltage across the capacitor

$$e = \frac{q}{C} = \frac{S_q}{C}\, m = S_v\, m$$

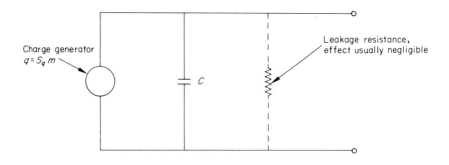

Fig. 1.31   Electrical equivalent of piezoelectric transducer

where $S_v = \dfrac{S_q}{C}$ is the so called voltage sensitivity of the transducer. An alternative equivalent representation of a piezoelectric transducer is that of a capacitance in series with a variable voltage source of a magnitude proportional to the measurement variable, figure 1.32.

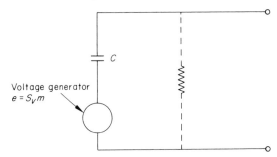

Fig. 1.32   Alternative electrical equivalent of piezoelectric transducer

A capacitive microphone consists of a variable capacitance in series with a fixed d.c. voltage source and has the equivalent circuit shown in figure 1.33. The measurement variable in this case produces proportional changes in capacitance $\Delta C = S_c\, m$, where S_c represents the sensitivity of the transducer. A change in capacitance is accompanied by a change in the charge on the capacitor so that the measurement problem is essentially the same as for the piezoelectric transducer.

Charge measurement can be performed by an operational current integrator; the principle is illustrated by the circuit shown in figure 1.34. The transducer is connected to the virtual earth at the amplifier summing point, any charge developed on the transducer capacitance is immediately, (neglecting R_1), forced onto the feedback capacitor C_f and the amplifier develops an output voltage

$$e_0 = -\frac{q}{C_f} \quad \text{(assuming } C_f \text{ initially uncharged).}$$

In the case of a piezoelectric transducer

$$e_0 = -\frac{S_q}{C_f} m = S_v \frac{C}{C_f} m \quad \text{and for a variable capacitance}$$

transducer a change in capacitance ΔC causes an output voltage

$$e_0 = -\frac{\Delta q}{C_f} = -E\frac{\Delta C}{C_f} \quad \text{(assuming } C_f \text{ initially uncharged).}$$

Linear Integrated Circuit Applications

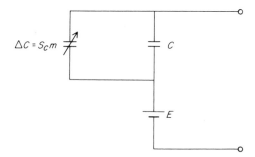

Fig. 1.33   Electrical equivalent of capacitive microphone

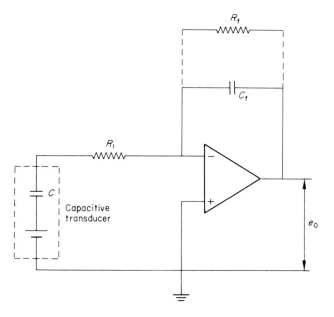

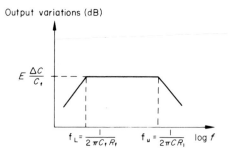

Fig. 1.34   Charge amplifier for capacitive transducer

Any external capacitance in parallel with the transducer, this includes cable capacitance, is connected between the virtual earth at the amplifier summing point, and ground. It is uncharged and does not, therefore, affect the measurement. The system is thus relatively unaffected by changes in cable length between the transducer and the amplifier, provided that the cable behaves like a capacitance. If connecting cables are excessively long, high frequency pulses generated by the transducer can be reflected back to the transducer causing serious signal distortion. A working rule is that a coaxial cable acts like a capacitance if the length of the cable is less than 1/40 of the wavelength in the cable of the highest frequency of interest. The connection cable should be no longer than

$$L = 1/40 \frac{KC}{f}$$

$L =$ length in metres
$K =$ propagation constant (0.66 for coaxial cable)
$C =$ velocity of light in metres/sec.

Substitution of numerical values indicates that this limitation of cable length is not usually of significance even at the highest frequencies at which capacitive transducers are designed to operate.

In a practical charge amplifier it is necessary to provide a d.c. path in the form of a resistor R_f connected between the amplifier output terminal and the inverting input terminal, otherwise amplifier bias current causes a continual charging of capacitor C_f and the output drifts into saturation. The presence of the resistor R_f limits the lower bandwidth limit of the charge amplifier to a frequency

$$f_L = \frac{1}{2\pi C_f R_f}$$

and transducing of very low frequency variations of the measurement variable requires a large value for the time constant $C_f R_f$. The gain of the charge amplifier varies inversely with the value of C_f, this rules out a large value for C_f and requires the use of a very large value for R_f. The use of very large values for R_f requires an operational amplifier type with very low bias current in order that offset and drift error should not be excessive. An F.E.T. amplifier is indicated or, for the ultimate in low bias current drift, a varactor bridge amplifier. The resistor R_1 in figure 1.34 is connected in series with the transducer in order to make the feedback loop stable, it sets the upper bandwidth limit of the charge amplifier at a frequency

$$f_u = \frac{1}{2\pi C R_1}$$

An alternative method of conditioning the signal from a capacitive transducer is to use a very high input impedance voltage amplifier. The main operational difference between the voltage amplifier approach and the charge amplifier approach is that the use of a voltage amplifier makes the system sensitive to external capacitance in parallel with the transducer. Changes in the length of the connecting cable between a capacitive transducer and voltage amplifier have a marked effect on the system performance.

Consider the circuit shown in figure 1.35 in which an operational amplifier configured as a high input impedance follower with gain, is used to condition the signal from a piezoelectric transducer. Capacitance C_2 represents capacitance to ground at the non-inverting input terminal and includes connecting cable capacitance, the resistance R_3 is required to provide amplifier bias current. If the magnitude of the transducer equivalent voltage generator varies sinusoidally the input signal to the amplifier is determined by the relationship

$$e_i = e \frac{C_1}{C_1 + C_2} \frac{1}{1 + \frac{1}{j\omega(C_1+C_2)R_3}}$$

and the amplifier output signal is

$$e_o = (1 + \frac{R_2}{R_1}) e_i$$

Note that at frequencies less than $f_L = \frac{1}{2\pi(C_1 + C_2)R_3}$

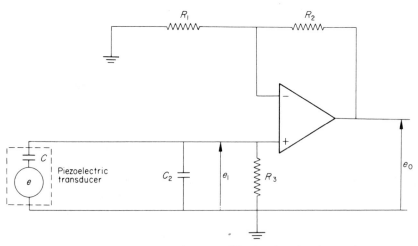

Fig. 1.35   High imput impedance amplifier for piezoelectric transducer

the amplifier gives an output signal proportional to the first derivative of the input variable, f_L represents the lower bandwidth limit of the system. The upper bandwidth limit is determined by the amplifier open loop response and dependent upon the frequency content of the input variable, it may be desirable to restrict the bandwidth by connecting a capacitance across R_2. Provision of an adequate low frequency response is often a design problem and requires a very large value for resistor R_3. The magnitude of R_3 that may be used is influenced by accuracy requirements as set by amplifier bias current and its drift; the use of a low bias current amplifier, such as an F.E.T. input type, is favoured. Bootstrapping of the bias resistor may be usefully employed in order to avoid the use of excessively large resistors, the technique illustrated by the circuit in figure 1.36. In this circuit the input resistance measured at the non-inverting input terminal of the amplifier is effectively the common mode input impedance of the amplifier. If an amplifier type with a very large common mode input impedance is used, the low frequency bandwidth of the system will be determined by the time constant $C_1 R_1$.

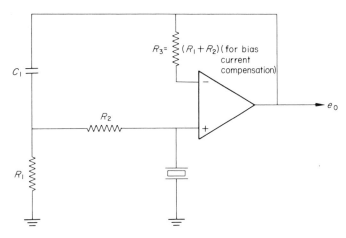

Fig. 1.36 Very high imput impedance amplifier for piezoelectric transducer

1.8 Voltage and Current Meter Circuits

Meters used to measure voltages and currents in a circuit should ideally not affect the quantity that is being measured. In order to satisfy this requirement a voltage measuring device should have a very high effective input impedance (ideally infinite), and a current measuring device a very low input impedance (ideally zero). Most pointer-scale meters use some form of moving coil meter movement as a basic indicating device. A moving coil meter is a current sensitive device, in that pointer deflection is proportional to current flow through the meter, and the meter has a non-zero resistance due to the resistance of the fine copper wire used to wind the coil.

A 0.1 mA meter movement may have a resistance of typically 1kΩ and an accuracy and resolution of about 1 per cent of full scale. If the meter is used to measure current its 1kΩ resistance will clearly affect the current being measured, or if it is scaled as a voltmeter (with appropriate series resistance) it will impose a 0.1mA load at the measurement points which will change the voltage which is being measured. By combining the moving coil meter with an operational amplifier circuit it is possible to reduce the effect that the meter has on the measurement circuit to negligible proportions and so achieve measurement accuracies of the same order as the resolution of the meter movement.

1.8.1 *D.C. Voltage Measurements*

An obvious approach to d.c. voltage measurements, illustrated in figure 1.37 is to use an operational amplifier configured as a high input impedance unity gain follower in order to prevent the meter loading the test circuit. The value of the series resistor R is chosen for voltage scaling; it should be a stable resistor with a low temperature coefficient of resistance. A 0.1 mA movement with resistance 1kΩ requires a 99kΩ series resistor for a 10 volt full scale reading; a stable series resistor will clearly swamp out the temperature

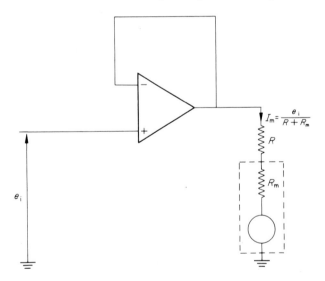

Fig. 1.37   Follower used for high resistance voltmeter

dependence of the resistance of the meter movement. Sensitivity can be increased by connecting the operational amplifier as a follower with gain.

An alternative approach is to use a circuit configuration in which the current through the meter movement is independent of the resistance of the meter movement, the circuit is illustrated in figure 1.38. In this circuit, negative feedback forces the input terminals of the amplifier to be at the same potential and causes a current $I_m = e_i/R$ to flow through the meter. The circuit has the high input impedance characteristic of the follower configuration.

Only voltage measurements with respect to ground are allowed by the two circuits discussed above. The circuit of figure 1.37 can be modified for differential measurements by adding a second unity gain follower used to drive the lower end of the meter circuit.

An alternative arrangement is to drive the meter movement and its series resistor with the two-amplifier differential circuit of figure 1.6, or a fully-developed instrumentation amplifier can be used for voltage measurements as discussed in section 1.4. Another differential circuit which has the advantage of a meter current independent of meter resistance is shown in figure 1.39, or if battery powering is allowable the circuit of figure 1.38 can be used as a floating differential voltmeter as shown in figure 1.40.

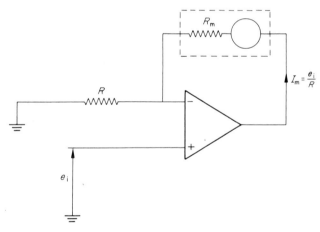

Fig. 1.38   High resistance voltmeter circuit

1.8.2 D.C. Current Measurements

Current measurements made with operational amplifier meter combinations are based upon the current to voltage converter configuration. A direct reading

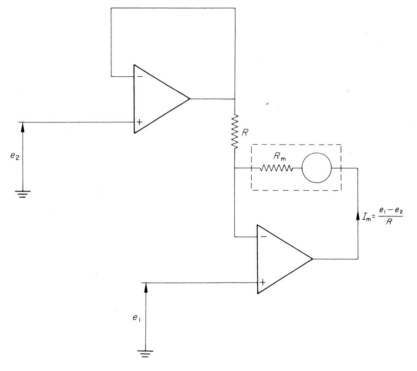

Fig. 1.39   High resistance differential voltmeter

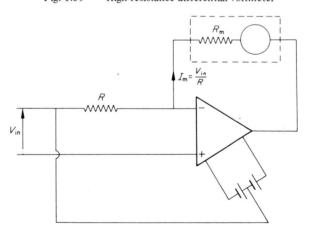

Fig. 1.40   Battery powered floating voltmeter

current meter circuit is illustrated in figure 1.41. The important point to realise is that the current measurement is made under almost perfect short circuit conditions, the amplifier forces all current arriving at its summing point to flow through the feedback path and maintains a virtual ground at the inverting input terminal. Operational amplifiers have open loop gains of typically 80 dB and upwards so that even for an amplifier output voltage of 10 volts, the voltage disturbance at the input is at the most only 1 millivolt.

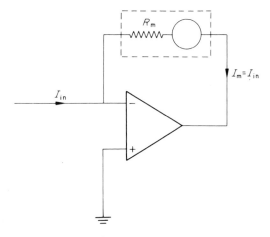

Fig. 1.41   Direct reading, zero voltage drop, current meter

It is not difficult to increase the current measurement sensitivity to a value many times greater than that of the basic meter movement. The principle is illustrated in figure 1.42. A resistive divider forces the amplifier to drive a current through the meter which is greater than the current to be measured by the factor $(1 + R_2/R_1)$. The sensitivities attainable with this type of circuit configuration are limited by accuracy requirements as dictated by amplifier bias current and its temperature drift. Operational amplifiers of the F.E.T. input variety should be used for the measurement of very small currents or, for extreme sensitivity and accuracy, a varactor bridge type of amplifier can be used. Note that current measurement circuits should not be used to measure current from a low source resistance. The magnitude of the source resistance together with the effective feedback resistance determines the voltage feedback fraction

$$\beta = \frac{R_s}{R_f + R_s}$$

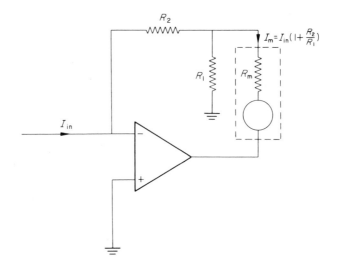

Fig. 1.42   Zero voltage drop current meter with increased sensitivity

A small value of R_s causes a small value for β and can lead to inadequate loop gain and an excessive gain error factor.

The current measurement circuits discussed measure current to ground. The differential input voltage amplifier configurations discussed earlier in this chapter can be used for a non-grounded current measurement but only through the use of a current sampling resistor and its consequent intrusion on the measurement circuit, a non-grounded, zero volts drop current measurement, can be made with both the circuits of figure 1.41 and 1.42 if the whole measurement circuit is floated and battery operation is employed. The technique is illustrated by the circuit shown in figure 1.43.

1.8.3 A.C. Measurements

Those d.c. measurement circuits, discussed above, in which the meter current is independent of meter resistance can all be modified to allow average reading a.c. operation. The modification consists of replacing the meter with a meter/ diode bridge arrangement and is illustrated in figure 1.44 for the case of the zero voltage drop current measuring circuit. Note that the non-linearity associated with the diodes is reduced to negligible proportions by including them within the feedback loop. Bandwidth limitations of the a.c. measurement circuits are determined by the slowing rate capabilities of the operational amplifier used in the circuits.

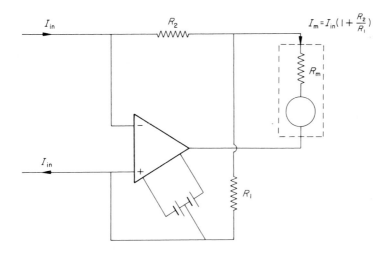

Fig. 1.43   Battery operated floating current meter

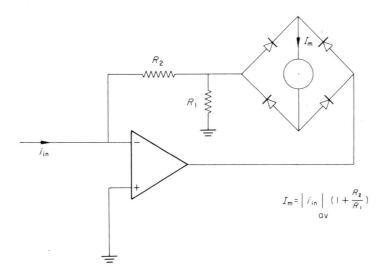

Fig. 1.44   A.C. current meter

Exercises 1

The exercises involving operational amplifiers assume a knowledge of the significance of operational amplifier characteristics and their effect on circuit performance errors. Such a knowledge will be obtained from a reading of the author's earlier book. Some general guidance notes are now given.

General Notes for Guidance

In negative feedback circuit configurations the open loop gain of the amplifier used in the circuit controls the closed loop gain errors and the small signal open loop frequency response characteristics of the amplifier directly influence the small signal closed loop bandwidth and the closed loop stability.

Gain Terminology

It is important to distinguish between the several 'gain' terms which are often used when discussing operational amplifier feedback circuits.

Open Loop Gain A_{OL}, may be defined as the ratio of a change of output voltage to the change in the input voltage which is applied directly to the amplifier input terminals.

The other gains are dependent upon both the amplifier and the circuit in which it is used and are controlled by the feedback fraction β.

The feedback fraction, β is the fraction of the amplifier output voltage which is returned to the input, it is a function of the entire circuit from output back to input, including both designed and stray circuit elements and the input impedance of the amplifier.

Loop Gain, βA_{OL} is the total gain in the closed loop signal path through the amplifier and back to the amplifier input via the feedback network. The magnitude of the loop gain, is of prime importance in determining how closely circuit performance approaches the ideal. The magnitude/phase relationships of βA_{OL} control closed loop stability.

Closed Loop Gain This is the gain for signal voltages connected directly to the input terminals of the amplifier. The closed loop gain for an ideal amplifier circuit is $1/\beta$ and for a practical circuit is

$$\frac{1}{\beta} \left[\frac{1}{1 + \dfrac{1}{\beta A_{OL}}} \right]$$

The quantity $\dfrac{1}{1 + \dfrac{1}{\beta A_{OL}}}$ is called the gain error factor the amount by which this factor differs from unity represents the gain error (usually expressed as a percentage).

Signal Gain, is the closed loop transfer relationship between the output and any signal input to an operational amplifier circuit.

Care should be taken to avoid confusion between closed loop gain and signal gain, in some circuits, the follower for example, the two gains are identical. Reference to the circuit shown in figure E.1.1 should clarify the distinction between the two types of gain.

It is worth noting that amplifier offset voltage, drift and noise, as specified, are referred to the input and are represented in terms of equivalent generators at the input connected at the position occupied by the signal source e_3 or e_4 in figure E.1.2. They appear at the output multiplied by the closed loop gain which for βA_{OL} much greater than 1 is equal to $1/\beta$. It is for this reason that $1/\beta$ is referred to as the 'noise' gain. An amplifier input offset voltage e_{os} becomes $1/\beta\, e_{os}$ at the output but when referred to the signal inputs in figure E.1.1 it becomes

$$e_{os}\, \frac{1}{\beta}\, \frac{R_1}{R_f}$$

at the e_1 input and $e_{os}\, \dfrac{1}{\beta}\, \dfrac{R_2}{R_f}$ at the e_2 input.

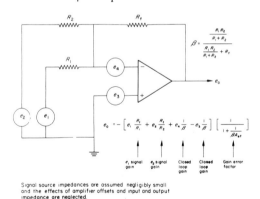

Fig. E.1.1 Difference between signal gain and closed loop gain

The General Offset and Drift Case

An appreciation of the difference between closed loop gain and signal gain, may be used to set out a general method for characterising and evaluating

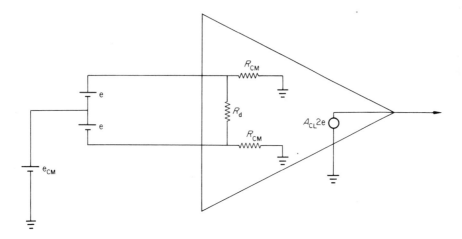

Fig. E.1.2 Equivalent circuit for one-amplifier differential circuit of figure 1.1

offset and drift errors.

Express all offsets in terms of equivalent voltages connected directly to the input terminals. Thus I_{b-} applies a voltage $-I_{b-} R_{-\text{source}}$ to the inverting input terminal and I_{b+} applies a voltage $-I_{b+} R_{+\text{source}}$ to the non-inverting input terminal.

$R_{\text{source}-}$ and $R_{\text{source}+}$ represent the effective source resistances connected at the inverting and non-inverting input terminals respectively. They represent the parallel combination of all resistive paths to ground, including, in the case of $R_{\text{source}-}$, the path through any feedback resistor and the amplifier output resistance to ground.

V_{io} is directly applied to the input terminal so we may represent the total equivalent input offset voltage as

$$E_{os} = \pm V_{io} + I_{b-} R_{\text{source}-} - I_{b+} R_{\text{source}+}$$

Drift is the total equivalent input offset voltage and is obtained by substituting values of the drift coefficients of I_b and V_{io}.

Graphs showing the dependence of E_{os} drift on source resistance are given in some amplifier data sheets. E_{os} appears at the output multiplier by the 'noise gain' $1/\beta$, the resultant error may be referred to any signal input by simply dividing by the signal gain associated with that input.

Amplifier Parameters

The characteristics of the operational amplifier to be used in the numerical examples are

Open loop gain 80 dB
First Order frequency response determined by

$$A_{OL}(j\omega) = \frac{10^4}{1 + j\,10^{-3}\omega}$$

CMRR = 80 dB I_B = 500 nA.

$\dfrac{\Delta I_B}{\Delta T}$ = 1 nA/°C, I_{io} = 50 nA, $\dfrac{\Delta I_{io}}{\Delta T}$ = 0.1 nA/°C

V_{io} = 2 mV, $\dfrac{\Delta V_{io}}{\Delta T}$ = 10 μV/°C

References

1. G.B. Clayton, *Operational Amplifiers,* Butterworth (1971)

Questions

1.1 Show that for a balanced differential input signal the one-amplifier differential circuit of figure 1.1 may be represented by the equivalent circuit shown in figure E.1.2. where the differential input resistance $R_d = 2R_1$, $R_{cm} = R_1 + R_2$ and $A_{CL} = R_2/R_1$. Assume that the operational amplifier in figure 1.1 behaves ideally.

1.2 Assuming that the operational amplifier used in the circuit behaves ideally, derive an expression for the CMRR of the differential feedback configuration, figure 1.1, due to resistor tolerance. Assume that resistor R_1 becomes $R_1 \pm K$ and R_2 becomes $R_2 \pm K$ where K is the resistor tolerance. Assume a worst case distribution of resistor values.

1.3 A one-amplifier differential circuit (figure 1.1) is to be given a differential input resistance of 100kΩ and a closed loop gain of 10. Find the values required for the resistors R_1 and R_2. If the initial offsets are balanced and the amplifier used in the circuit has the drift characteristics given in Exercises 1 what is the minimum differential input signal that can be amplified with less than 1 per cent error due to a temperature change of 10°C?

1.4 Show that the CMRR of the one-amplifier differential circuit due to the combined effects of resistor tolerance and amplifier CMRR is

$$\text{CMRR} = \frac{\text{CMRR}_A \times \text{CMRR}_R}{\text{CMRR}_A + \text{CMRR}_R}$$

Where CMRR_A is the common mode rejection ratio of the operational amplifier used in the circuit and CMRR_R is the common mode rejection ratio of the circuit due to resistor tolerance.

1.5 Derive the ideal performance equation for the circuit of figure 1.4.

1.6 Using operational amplifiers with ± 10 volts output capability, sketch a two-amplifier differential circuit with differential input resistance 40kΩ differential gain 10 and with a maximum common mode range of ± 100 volts. What will be the input resistance of the circuit to common mode signals?

1.7 Show that the CMRR of the differential input amplifier in figure 1.6 is

$$\text{CMRR} = \frac{\text{CMRR}_{A_1} \times \text{CMRR}_{A_2}}{\text{CMRR}_{A_1} + \text{CMRR}_{A_2}}$$

If $\text{CMRR}_{A_1} = 80$ dB' and $\text{CMRR}_{A_2} = 72$ dB what are the possible values for the CMRR of the circuit?

1.8 A differential input signal of 200 mV, is to be used to drive a constant current of 2 mA through an earthed load resistor R_L. The maximum value of R_L is 3kΩ but it may take on any value less than 3kΩ.

(a) Choose the component values to be used in the circuit of figure 1.13. The circuit is to have a differential input resistance of 20kΩ. Assume that the operational amplifier in the circuit has a maximum output voltage capability of ± 10 volts but that it behaves ideally in all other respects.

(b) What will the CMRR of the circuit be due to resistor mismatch caused by loading at the reference input?

(c) What advantages would there be in using the AD 520 instrumentation amplifier instead of the one-amplifier differential circuit?

1.9 A differential input instrumentation amplifier has CMRR 80 dB when supplied by a balanced source. If $R_{cm} = 10^7$ ohms what is the CMRR of the circuit when supplied by a single-ended signal source of resistance 10kΩ (see figure 1.19).

1.10 In figure 1.23 $R_3 = R_4 = R_1 = R_0 = 100$kΩ and $R_2 = R_0(1 + \alpha)$, $E = 10$ volts. What is the smallest percentage change in R_2 that can be measured with less than 1 per cent error due to a temperature change of 10°C. Assume that initial offsets are balanced and that the operational amplifier used in the circuit has the temperature drift characteristics given in Exercises 1.

1.11 The circuit of figure 1.38 is to be used as the basis of a high resistance voltmeter to give a full scale deflection for 10mV. If a 0.1 mA meter movement of resistance 1kΩ is used what value is required for the resistor R? What is the error due to a temperature change of 10°C? Assume initial amplifier offset is balanced out and that the amplifier used in the circuit has the temperature drift characteristic given in Exercises 1.

1.12 A 0.1 mA 1kΩ resistance meter movement is to be used in the circuit of figure 1.42 to read a current of 1 μA. What value should be used for the ratio R_2/R_1? What is the error due to a temperature change of 10°C. Assume initial amplifier offset is balanced and that the amplifier used in the circuit has the temperature drift characteristics given in Exercises 1.

2. Some Signal Processing Applications

In a book of this nature the division of the subject matter under separate chapter headings must at times be somewhat arbitrary. In this chapter we gather together some particular signal processing applications in which linear integrated circuits can be used to advantage, many of the circuits presented in other chapters of the book can equally well be regarded as signal processing applications.

2.1 Active Filters using Operational Amplifiers

In electronic systems the desired signal is often obscured by undesired signals. Signals are distinguished by their frequency characteristics and the extraction of a discreet signal from unwanted signals can be accomplished by some form of frequency selective circuit or filter.

Signal filtering can be performed by circuits which use only passive components, resistors, capacitors, inductors, but so called active filters, which use an active gain element, (conveniently an op. amp.), together with a resistor-capacitor network offer considerable advantages. They eliminate the need for inductors and inductors are relatively large, heavy and costly elements, particularly those for use at frequencies in the audio range and below. In addition active filters, dependent upon the circuit configuration adopted, can provide gain and excellent isolation properties (high input impedance, low output impedance).

The theory underlying the design of active filters has received extensive coverage in the literature [1,2,3], e.g., low pass, high pass, band pass and band reject filters can all be realised, each by means of a variety of different circuit configurations. The decision as to which is the best circuit to use for a specific filter function is very much dependent upon the detailed design objectives, logically the selection criterion should rest on a cost/performance comparison between the different possible circuit configurations. Unfortunately, if the designer is not thoroughly familiar with active filter circuits he could well spend a great deal of time in making such a comparison and this would completely cancel any possible cost benefits if his requirements were for only a small number of filter circuits.

Linear Integrated Circuit Applicstions

The treatment of active filter circuits to be given here is not intended as a survey of possible filter realisations, a limited number of designs have been selected for discussion, the circuits work and will be found suitable for a wide range of filter applications. The performance parameters of active filters are very much dependent upon the values of the resistors and capacitors used in the circuits and upon the temperature stability of these components. Precision resistors and capacitors are expensive and trimming component magnitudes to precise values is a time consuming process, for these reasons the circuits presented are, in the main, those which require a minimum number of passive components. The discussion of the filter circuits which is to be given assumes a familiarity with transfer function terminology, (See Appendix A).

2.2 Low Pass Active Filters

1st Order A first order low pass filter can be made quite easily with a single resistor and capacitor but the inclusion of an operational amplifier as an active element gives the circuit a low output impedance and makes it insensitive to loading. A circuit for a first order low pass filter is given in figure 2.1. Ideally the filter has the transfer function

$$A_{(s)} = \frac{R_2}{R_1} \frac{1}{1 + CR_2\, s} \qquad (2.1)$$

with a steady state sinusoidal response of magnitude

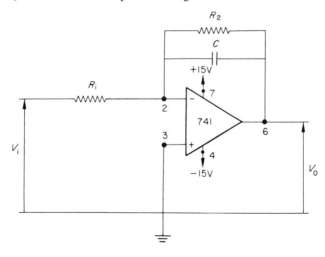

Fig. 2.1   First order low pass filter

$$\left[A(j\omega)\right] = \frac{R_2}{R_1} \frac{1}{\left[1 + \left(\frac{\omega}{\omega_0}\right)^2\right]^{1/2}} \tag{2.2}$$

where $\omega_0 = \dfrac{1}{CR_2}$

Performance errors may arise because of inadequate loop gain, and as a result of amplifier drift[4].

Second Order Our treatment of second order low pass active filters is limited to those filter realisations in which the active element (the op. amp.), acts as a so called voltage controlled voltage source, VCVS. An ideal VCVS is a voltage amplifier with infinite input impedance, zero output impedance and which gives an output voltage equal to the input voltage multiplied by some constant coefficient. The high input impedance, low output impedance and stable gain of an operational amplifier, when connected in the non-inverting feedback configuration, allows its use as a close approximation to the ideal VCVS. The VCVS low pass filter realisation has been chosen for treatment because this circuit arrangement, as we shall see, allows for independent adjustment of the transfer function parameters, ω_0 and ζ, and because it requires a comparatively small number of passive components, VCVS filter realisations have the disadvantage of giving transfer function parameters which are very sensitive to component value changes if small values of the damping factor and large values of gain constant (> 10) are required. However, this does not present a serious problem in the low pass filter case, because the low pass filter functions which are used in practice do not generally call for very small values of the damping factor and if gain is required it is comparatively simple and inexpensive to use a second operational amplifier as a separate gain stage.

A VCVS filter circuit for the realisation of a second order low pass transfer function is given in figure 2.2. The transfer function for the network may be put in the form given in equation A.7. as

$$\frac{V_{o(s)}}{V_i} = \frac{A_0}{1 + b\dfrac{s}{\omega_0} + \dfrac{s^2}{\omega_0^2}} \tag{2.3}$$

where the transfer function parameters are determined by circuit values according to the relationships

$$A_0 = K = 1 + \frac{R_4}{R_3} \quad \text{the gain of the VCVS}$$

$$\omega_0 = \left(\frac{1}{R_1 R_2 C_1 C_2}\right)^{1/2}$$

$$b = \left(\frac{R_2 C_2}{R_1 C_1}\right)^{1/2} + \left(\frac{R_1 C_2}{R_2 C_1}\right)^{1/2} + \left(\frac{R_1 C_1}{R_2 C_2}\right)^{1/2} - K\left(\frac{R_1 C_1}{R_2 C_2}\right)^{1/2}$$

A filter design requires a choice of component values to give specified values for the transfer function parameters. In most filter designs it is usual to start by selecting capacitor values because there are fewer standard values of capacitors than there are resistors.

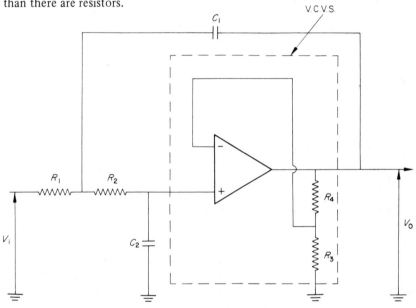

Fig. 2.2   VCVS low pass filter

Design Procedure 1 For a specified A_0, ω_0, b ($A_0 > 2$)

Make $C_1 = C_2 = C$ a convenient value.
Choose values of R_3 and R_4 to make

$$A_0 = K = 1 + \frac{R_4}{R_3}$$

Calculate the value required for R_2 from

$$R_2 = \frac{b}{2\omega_0 C} \left[1 + \left(1 + \frac{4(A_0 - 2)}{b^2}\right)^{1/2}\right]$$

Calculate the value required for R_1 from

$$R_1 = \frac{1}{\omega_0^2 C^2 R_2}$$

Example: A VCVS low pass filter is required with

$A_0 = 6$, $f_0 = 200$ Hz, $b = 0.6$

Calculations:

$\omega_0 = 2\pi f_0 = 1256$

We choose $C_1 = C_2 = 0.047\mu F$

$$R_2 = \frac{0.6}{2 \times 1256 \times 0.047 \times 10^{-6}} \left[1 + \left(1 + \frac{16}{0.36}\right)^{\frac{1}{2}}\right]$$

$= 39.36$ kΩ

$$R_1 = \frac{1}{1256^2 \times 0.047^2 \times 10^{-12} \times 39.36 \times 10^3}$$

$= 7.29$kΩ

$A_0 = K = 6 = 1 + \dfrac{R_4}{R_3}$

and $\dfrac{R_4}{R_3} = 5$

A potentiometer or selected resistor values may be used to fix this ratio. The practical circuit together with its frequency response is shown in figure 2.3. Resistor values were obtained by trimming. Note that ω_0 may be adjusted by changing R_1 and R_2 by equal percentages and this does not affect the value of b. A required value for b may be obtained by adjusting K without affecting the value of ω_0. Inspection of the second order low pass magnitude response curves given in figure A.1. shows that at a frequency $10\ \omega_0$ all curves are 40 dB down on their value at $0.1\ \omega_0$. This provides a generally applicable method of adjusting ω_0 for second order low pass filters; the appropriate ω_0 determining elements are adjusted so as to make the magnitude of the filter output, at a frequency $10\ \omega_0$, 40 dB down on the value at frequencies below $0.1\ \omega_0$.

Design Procedure 2 A simpler design procedure is possible if A_0 is allowed to be a free parameter, this is not a severe restriction for scaling can usually be performed elsewhere.

Choose $C_1 = C_2 = C$ a convenient value.
Make $R_1 = R_2 = R$ and find the value of R from

$$R = \frac{1}{\omega_0 C}$$

Find the value required for K from

$$b = 3 - K$$

and adjust K to its required value by choice of the ratio

$$\frac{R_4}{R_3}$$

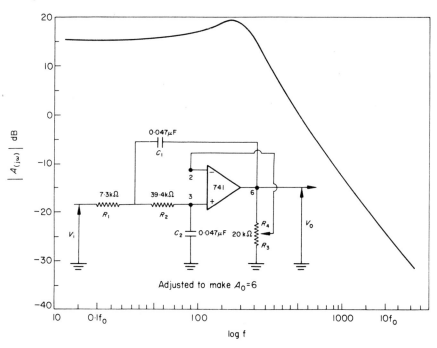

Fig. 2.3   VCVS low pass filter design; design procedure 1.

Design Procedure 3 Connect the operational amplifier as a unity gain follower making $K = A_0 = 1$

Choose $C_1 = C$ and $C_2 = \dfrac{C}{M}$ to be convenient values.

The value for M must be such that $\dfrac{M b^2}{4} \geqslant 1$

Calculate the value of R_2 from

$$R_2 = \frac{1}{\omega_0 C} \left[\frac{bM}{2} + \left(\frac{(bM)^2}{4} - M \right)^{\frac{1}{2}} \right]$$

Calculate the value of R_1 from

$$R_1 = \frac{M}{\omega_0^2 C^2 R_2}$$

Example: A second order low pass filter is required with

$f_0 = 500\ H_z,\ b = \sqrt{2}$ (Butterworth response)

Value of M must be chosen so that

$$\frac{M b^2}{4} \geqslant 1 \quad \text{that is } M \geqslant 2$$

Use $M = 2$

Make $C_1 = C = 0.02\ \mu F$ and $C_2 = \dfrac{C}{M} = 0.01\ \mu F$

$$R_2 = \frac{2\sqrt{2}}{2 \times 2\pi \times 500 \times 0.02 \times 10^{-6}} = 22.5\ k\Omega$$

$$R_1 = \frac{2}{4\pi^2 \times 500^2 \times 0.02^2 \times 10^{-12} \times 22.5 \times 10^3}$$

$= 22.5\ k\Omega$

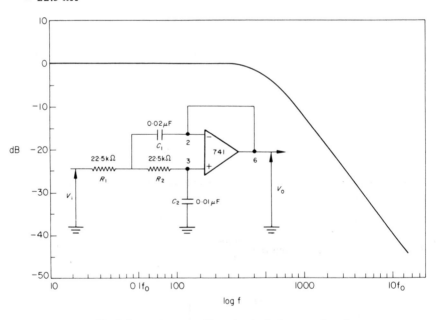

Fig. 2.4 Low pass filter circuit; design procedure 3

The practical filter circuit together with its frequency response is shown in figure 2.4.

Note: In the designs considered, the choice of capacitor value determines the resistor values required for a particular value of ω_0. Small value close tolerance capacitors are less expensive than the larger values but smaller capacitor values require larger resistor values with a greater possible error due to drift in amplifier bias current.

2.3 High Pass Active Filters

1st Order A first order high pass filter, like the first order low pass, can be made simply with a resistor and a capacitor but the addition of an operational amplifier gives the circuit a low output impedance. A first order high pass active filter circuit is shown in figure 2.5. Ideally the circuit has the transfer function

$$A(s) = \frac{\frac{R_2}{R_1} s}{s + \frac{1}{C_1 R_1}} \qquad (2.4)$$

with a steady state sinusoidal response of magnitude

$$\left[A(j\omega)\right] = \frac{\frac{R_2}{R_1}}{\left(1 + \left(\frac{\omega_0}{\omega}\right)^2\right)^{\frac{1}{2}}} \qquad (2.5)$$

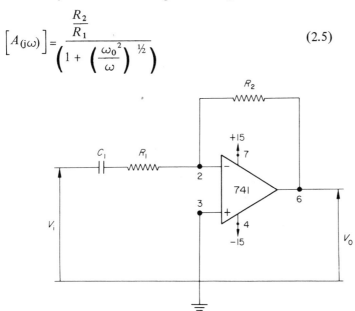

Fig. 2.5   First order high pass filter

2nd Order Again we choose to present a VCVS realisation and in figure 2.6 we show a circuit which gives a second order high pass transfer function. Comparing the circuit with the low pass circuit of figure 2.2 we see that the positions of the resistors and capacitors in the circuit have been interchanged. The transfer function for the filter in figure 2.6 may be put in the form given in the Appendix, equation A.22.

$$A(s) = \frac{A_0 s^2}{s^2 + b s \, \omega_0 + \omega_0^2} \tag{2.6}$$

with the constants, A_0, b and ω_0 for the response determined by component values according to the relationships

$$A_0 = K = 1 + \frac{R_4}{R_3}$$

$$\omega_0 = \left(\frac{1}{R_1 R_2 C_1 C_2}\right)^{1/2}$$

$$b = \left(\frac{R_1 C_1}{R_2 C_2}\right)^{1/2} + \left(\frac{R_1 C_2}{R_2 C_1}\right)^{1/2} + \left(\frac{R_2 C_2}{R_1 C_1}\right)^{1/2} - K \left(\frac{R_2 C_2}{R_1 C_1}\right)^{1/2}$$

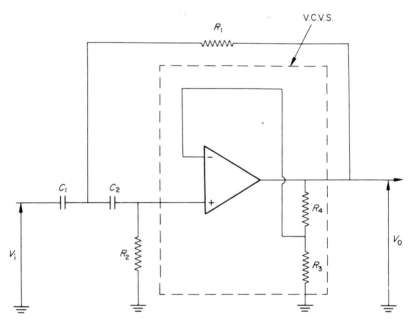

Fig. 2.6   VCVS high pass filter

The VCVS high pass filter realisation, in the same way as the low pass circuit, allows for independent adjustment of ω_0 and ζ; ω_0 can be adjusted by making equal percentage changes in R_1 and R_2, ζ can be adjusted by a control of K by means of the resistor ratio R_4/R_3. All second order high pass filter functions have a response magnitude at $0.1\,\omega_0$ which is 40 dB down on the magnitude at $10\,\omega_0$ and, as in the low pass case, this provides a convenient measurement guide when making practical adjustments in order to tune the filter to a particular value of ω_0.

Design Procedure 1 For a specified A_0, ω_0, b,

choose the resistor ratio $\dfrac{R_4}{R_3}$ to make $A_0 = K = 1 + \dfrac{R_4}{R_3}$

Make $C_1 = C_2 = C$, a convenient value.
Calculate the value required for R_1 from

$$R_1 = \frac{b + (b^2 + 8(A_0 - 1))^{1/2}}{4\,\omega_0\,C}$$

Calculate the value required for R_2 from

$$R_2 = \frac{1}{\omega_0^2\,R_1\,C^2}$$

Design Procedure 2 For a specified ω_0, b. With A_0 a free parameter
Make $C_1 = C_2 = C$ a convenient value.
Make $R_1 = R_2 = R$ and calculate the value required from

$$R = \frac{1}{\omega_0\,C}$$

Find the value required for K from

$$b = 3 - K$$

and adjust K to this value by means of the ratio R_4/R_3.

Design Procedure 3 For a specified ω_0 and b with $A_0 = 1$
Connect the operational amplifier as a unity gain follower making $K = 1$
Make $C_1 = C_2 = C$ a convenient value
Calculate the value required for R_2 from

$$R_2 = \frac{2}{b\,\omega_0\,C}$$

Calculate the value required for R_1 from

$$R_1 = \frac{b}{2\,\omega_0\,C}$$

Example. A second order high pass filter is required with $f_0 = 500$ Hz, $b = \sqrt{2}$ (Butterworth response)

Make $C_1 = C_2 = 0.01 \mu F$

$$R_2 = \frac{2}{\sqrt{2}\ 2\pi\ 500\ 0.01 \times 10^{-6}} = 45 k\Omega$$

$$R_1 = \frac{\sqrt{2}}{2\ 2\pi\ 500\ 0.01 \times 10^{-6}} = 22.5 k\Omega$$

The practical filter circuit together with its frequency response is shown in figure 2.7.

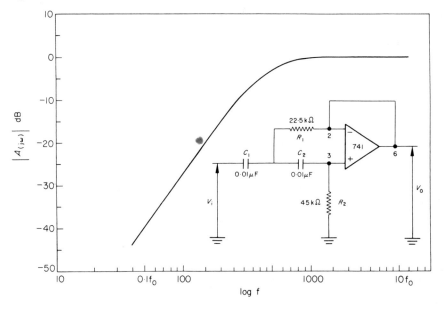

Fig. 2.7   High pass filter; design procedure 3

Note. The response of all the high pass circuit falls off at high frequencies because of the open loop bandwidth limitations of the operational amplifier. The effect may be understood in terms of the Bode plots shown in figure 2.8.

2.4 Band Pass Active Filters

2.4.1 *VCVS Band Pass Filter*

A VCVS circuit with a second order band pass transfer function is shown in

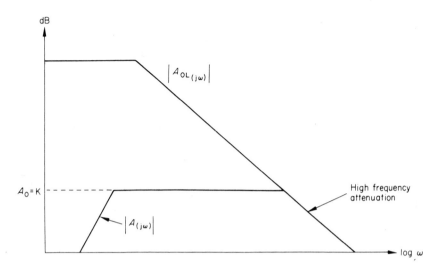

Fig. 2.8   Bode plot for high pass filter

figure 2.9. The voltage transfer function for the circuit may be put in the form given in the Appendix, (equation A.25)

$$A(s) = \frac{A_0 \, b \, \omega_0 \, s}{s^2 + b \, \omega_0 \, s + \omega_0^2} \tag{2.7}$$

The relationships which give the transfer function parameter in terms of the circuit parameters are somewhat cumbersome, they are

$$A_0 = \frac{K}{1 + \dfrac{R_1}{R_3} + \dfrac{C_2}{C_1}\left(1 + \dfrac{R_1}{R_2}\right) + (1-K)\dfrac{R_1}{R_2}}$$

$$\omega_0 = \left[\frac{1}{R_3}\left(\frac{1}{R_1} + \frac{1}{R_2}\right)\frac{1}{C_1 C_2}\right]^{1/2}$$

$$b = \frac{1}{Q} = \left(\frac{R_3}{\dfrac{1}{R_1} + \dfrac{1}{R_2}}\right)^{1/2} \left[\sqrt{\frac{C_1}{C_2}}\left(\frac{1}{R_1} + \frac{1}{R_3} + \frac{1-K}{R_2}\right) + \sqrt{\frac{C_2}{C_1}}\left(\frac{1}{R_1} + \frac{1}{R_2}\right)\right]$$

Design Procedure   Considerable simplification of the design equations can be obtained by setting $C_1 = C_2 = C$ and $R_1 = R_2 = R_3 = R$. A_0 must be a free parameter.

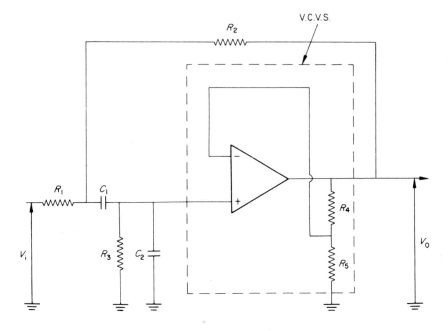

Fig. 2.9   VCVS band pass filter

A specified value of ω_0 is required and $b = 1/Q$.
Choose $C_1 = C_2 = C$ a convenient value

Calculate R from

$$R = \frac{\sqrt{2}}{\omega_0 C}$$

Calculate K from

$$K = 5 - \frac{\sqrt{2}}{Q}$$

Choose the ratio $\dfrac{R_4}{R_3}$ to give this required value of K

With the value of Q fixed

$$A_0 = \frac{5}{\sqrt{2}} Q - 1$$

Like the other VCVS filters the bandpass circuit permits independent adjustment of ω_0 and Q, ω_0 is adjusted by making equal changes in the resistors

R, Q is adjusted by altering K by means of the resistor value R_4/R_3. The VCVS bandpass circuit is not very suitable if a Q of greater than 10 is required, high Q circuits have transfer function parameters which are very sensitive to component value changes.

2.4.2 *A Negative Inmittance Converter Band Pass Filter*

A band pass filter circuit which is not as sensitive to element value changes as the VCVS circuit, and which is therefore satisfactory for higher Q values, is shown in figure 2.10. The circuit is based upon the use of an operational amplifier as a current inverter, a so called 'ideal current-inversion-negative-inmittance converter', INIC[1].

The action of the operational amplifier as a current inverter is readily analysed in terms of the concepts of an ideal operational amplifier.

An ideal operational amplifier draws no current at its input terminals. Thus

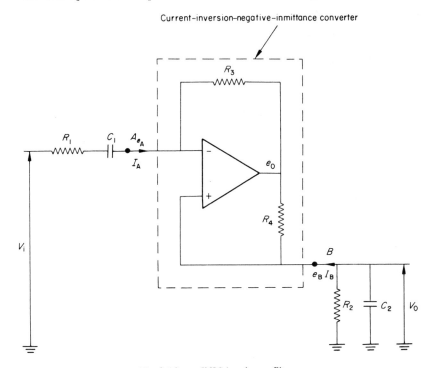

Fig. 2.10   INIC band pass filter

$$\frac{e_A - e_0}{R_3} = I_A$$

and $\dfrac{e_B - e_0}{R_4} = I_B$

Feedback forces the differential input terminals of the amplifier to be at the same potential. Thus

$$e_A = e_B$$

which makes

$$I_B = \frac{R_3}{R_4} I_A = G I_A$$

$G = \dfrac{R_3}{R_4}$ may be thought of as the 'gain' of the INIC.

If an impedance Z_B is connected between terminal B and earth then $e_B = -I_B Z_B$

and $\dfrac{e_B}{I_A} = \dfrac{e_A}{I_A} = -\dfrac{I_B}{I_A} Z_B = -G Z_B$

The input impedance measured between point A and earth is G times the negative of the impedance connected between point B and earth.

An operational amplifier current inverter circuit is not unconditionally stable since it includes both a positive and a negative feedback path. For stability the loop gain βA_{OL} must not be allowed to approach the value plus one. In the case of the current inverter

$$\beta = \frac{Z_B}{R_4 + Z_B} - \frac{Z_A}{R_3 + Z_A}$$

where Z_B is the impedance connected externally between point B and earth and Z_A is the impedance connected between point A and earth.

Returning to the INIC band pass circuit of figure 2.10 and using the fact that $I_B = G I_A$ it can be shown that the voltage transfer function for the circuit is

$$\frac{V_0}{V_i}(s) = \frac{G \dfrac{1}{R_1 C_2} s}{s^2 + s\left(\dfrac{1}{R_1 C_1} + \dfrac{1}{R_2 C_2} - \dfrac{G}{R_1 C_2}\right) + \dfrac{1}{C_1 C_2 R_1 R_2}} \qquad (2.8)$$

Equation 2.8 has the same form as the general second order band pass function which is given in the Appendix, (equation A 25). The transfer function parameter may be written in terms of the circuit parameter, as

$$A_0 = \frac{G}{\frac{C_2}{C_1} + \frac{R_1}{R_2} - G}$$

$$b = \frac{1}{Q} = \left(\frac{R_2 C_2}{R_1 C_1}\right)^{1/2} + \left(\frac{R_1 C_1}{R_2 C_2}\right)^{1/2} - G\left(\frac{R_2 C_1}{R_1 C_2}\right)^{1/2}$$

$$\omega_0 = \left[\frac{1}{C_1 C_2 R_1 R_2}\right]^{1/2}$$

Note that the filter output signal is *not* available at the low impedance output terminal of the operational amplifier but is produced at a high impedance point in the circuit. This represents a disadvantage of the INIC filter realisation when compared with the VCVS circuits. Satisfactory operation of the INIC circuit requires the use of an output buffer stage for isolation purposes.

Design Procedure For a specified ω_0 and Q with A_0 a free parameter.
Choose $C_1 = C_2 = C$ a convenient value.
Make $R_1 = R_2 = R$ and calculate the required value from

$$R = \frac{1}{\omega_0 C}$$

Calculate the value required for G from

$$G = 2 - \frac{1}{Q}$$ and set this value by choice of the resistor ratio $R_3/R_4 = G$.

Example. An INIC realisation of a second order band pass filter is required with $f_0 = 600$ Hz, $Q = 100$.
Choose $C_1 = C_2 = C = 0.01\ \mu F$

$$R = \frac{1}{\omega_0 C} = \frac{1}{2\pi\,600 \times 0.01 \times 10^{-6}} = 26.5\text{k}\Omega$$

$$\frac{R_3}{R_4} = G = 2 - \frac{1}{100} = 1.99$$

gives $A_0 = \frac{G}{2 - G} = \frac{1.99}{0.01} = 199$

A practical circuit together with its response is shown in figure 2.11. The potentiometer was used to trim the centre frequency gain to its value of 199. The response of the circuit with centre frequency gain trimmed to a value of 99, corresponding to a $Q = 50$, is also shown.

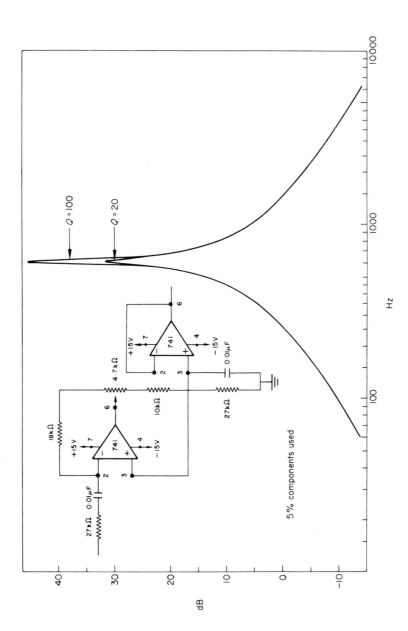

Fig. 2.11 Practical INIC band pass filter and its response

2.5 Filter Realisation Using Analogue Computer Techniques

A filter transfer operator can be regarded as a symbolic representation of a linear differential equation, (see Appendix A); equations of this type can be readily solved with operational amplifiers acting as integrators and adders and using conventional analogue computing techniques. We confine our discussion of the method to the realisation of second order filters but the technique may be readily extended to give higher order transfer functions.

Consider the second order low pass voltage transfer function (equation A.7).

$$\frac{V_{o(s)}}{V_i} = \frac{A_0}{1 + b\frac{s}{\omega_0} + \frac{s^2}{\omega_0^2}}$$

The complex variable s is replaced by the differential operator $p = \frac{d}{dt}$ and the equation is rearranged to give

$$\frac{p^2}{\omega_0^2} V_0 = A_0 V_i - \frac{b}{\omega_0} p V_0 - V_0 \qquad (2.9)$$

A differential equation of this form can be solved by the computing loop shown in figure 2.12. The loop consists of two integrators and an adder-subtractor. Expressed in terms of the differential operator p the action of an operational integrator is to multiply by

$$\frac{-1}{\tau p}$$

where τ is the 'characteristic time' of the integrator, determined by the product of capacitor-resistor values. For design simplicity the τ's for the two integrators are made the same and τ is used to represent $\frac{1}{\omega_0}$, with this representation the action of an integrator is to multiply by

$$\frac{-\omega_0}{p}$$

The action of the computing loops may be understood by assuming that a voltage $E_{01} = \frac{p^2}{\omega_0^2} V_0$ is available. E_{01} is applied as an input to the first integrator which produces an output signal $E_{02} = -\frac{p}{\omega_0} V_0$. This signal is integrated again by the second integrator to give $E_{03} = V_0$. An adder-subtractor amplifier is now used to form the voltage E_{01}. The resistor values in the adder-subtractor are chosen to simplify design procedure and in terms of these resistor values it produces an output signal

70 *Linear Integrated Circuit Applications*

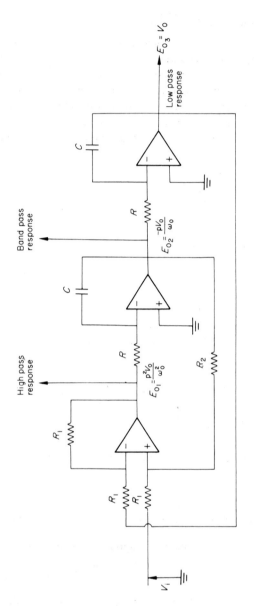

Fig. 2.12 Analogue computing loop used to realise filter transfer functions

Linear Integrated Circuit Applications 71

$$E_{01} = \frac{2}{R_2 + R_1} \left[V_i R_2 + E_{02} R_1 \right] - E_{03}$$

or

$$\frac{p^2}{\omega_0^2} V_0 = \frac{2 R_2}{R_2 + R_1} V_i - \frac{2 R_1}{R_2 + R_1} \frac{p V_0}{\omega} - V_0 \quad (2.10)$$

Low Pass Function The loop forces the voltages in the circuit to vary in the manner described by equation 2.10 and

$$\frac{V_0}{V_i}(p) = \frac{E_{03}}{V_i}(p) = \frac{2 \frac{R_2}{R_2 + R_1}}{1 + \frac{2 R_1}{R_2 + R_1} \frac{p}{\omega_0} + \frac{p^2}{\omega_0^2}} \quad (2.11)$$

a second order low pass transfer function. Comparison with the general equations allows the transfer function parameter to be written in terms of the circuit variables as

$$A_0 = 2 \frac{R_2}{R_2 + R_1}$$

$$b = 2 \frac{R_1}{R_2 + R_1}$$

$$\omega_0 = \frac{1}{CR}$$

Band Pass Function Remembering that $E_{02} = -\frac{p V_0}{\omega_0}$

equation 2.10 may be written as

$$-\frac{p}{\omega_0} E_{02} = 2 \frac{R_2}{R_2 + R_1} V_i + 2 \frac{R_1}{R_2 + R_1} E_{02} + \frac{\omega_0}{p} E_{02}$$

or

$$\frac{E_{02}}{V_i}(p) = -\frac{2 \frac{R_2}{R_2 + R_1} \omega_0 p}{p^2 + \frac{2 R_1}{R_2 + R_1} \omega_0 p + \omega_0^2} \quad (2.12)$$

equation 2.12 represents a second order band pass function; comparison with equation A.25 allows the transfer function parameters to be written in terms of the circuit variables as

$$A_0 = -\frac{R_2}{R_1}$$

$$b = \frac{2R_1}{R_2 + R_1}$$

$$\omega_0 = \frac{1}{CR}$$

High Pass Function Remembering that $E_{01} = \frac{p^2 V_0}{\omega_0^2}$

equation 2.10 may be written as

$$E_{01} = \frac{2R_2}{R_2 + R_1} V_i - \frac{2R_1}{R_2 + R_1} \frac{\omega_0}{p} E_{01} - \frac{\omega_0^2}{p^2} E_{01}$$

or

$$\frac{E_{01}}{V_i}(p) = \frac{2\frac{R_2}{R_2+R_1} p^2}{p^2 + 2\frac{R_1}{R_2+R_1} \omega_0 p + \omega_0^2} \qquad (2.13)$$

Equation 2.13 represents a second order high pass function with transfer function parameters related to circuit variables by the relationship

$$A_0 = 2\frac{R_2}{R_2 + R_1}$$

$$b = 2\frac{R_1}{R_2 + R_1}$$

$$\omega_0 = \frac{1}{CR}$$

The circuit of figures 2.12 simultaneously provides, low pass, high pass and band pass transfer functions. It requires three amplifiers and its use is therefore uneconomical for low Q, high pass and low pass functions. The circuit gives Q values which are less sensitive to component variations than those given by simple amplifier circuits and its use is therefore worth considering for high Q ($Q > 50$) band pass filters.

Design Procedure for Bandpass Realisation
ω_0 and Q are specified values.
Choose the capacitors to be of convenient values

Find R from

$$R = \frac{1}{\omega_0 C}$$

Determine the required value of the ratio $\frac{R_2}{R_1}$ from

$$\frac{R_2}{R_1} = 2Q - 1$$

Example. Required $f_0 = 600$ Hz, $Q = 50$
Make $C = 0.01\,\mu\text{F}$

$$R = \frac{1}{2\pi\, 600 \times 0.01} = 26.5\,\text{k}\Omega$$

$$\frac{R_2}{R_1} = 100 - 1 = 99$$

For the purpose of evaluation a circuit was connected together using components of 5 per cent tolerance. The circuit and its responses are shown in figure 2.13.

2.6 Band Reject Filters

Band reject filters are used to selectively attenuate a specific frequency or band of frequencies and ideally they produce no attenuation of frequencies outside the rejection band. The transfer function for a second order band reject network has the form

$$A_{(s)} = \frac{A_0\,(s^2 + \omega_0^2)}{s^2 + b\,\omega_0\,s + \omega_0^2} \tag{2.14}$$

with a steady state sinusoidal response of magnitude

$$|A(j\omega)| = \frac{A_0\,(\omega_0^2 - \omega^2)}{\sqrt{(\omega_0^2 - \omega^2)^2 + b^2\,\omega_0^2\,\omega^2}} \tag{2.15}$$

and phase angle

$$\phi = \tan^{-1} \frac{b\,\omega_0\,\omega}{\omega^2 - \omega_0^2} \tag{2.16}$$

The band reject transfer function is very similar to the band pass function, the parameter $Q = \omega_0/\omega_2 - \omega_1$ may be used as a measure of selectivity where ω_2 and ω_1 are the angular frequencies at which the magnitude response is 3 dB down on A_0. It can be proved by substitution and discarding negative solutions that $Q = \frac{1}{b}$

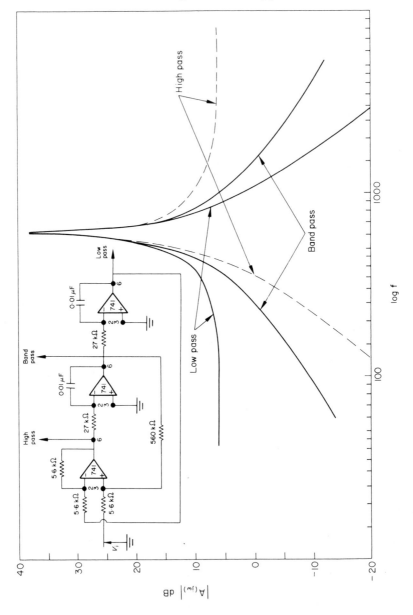

Fig. 2.13 Analogue computing loop used to realise filter transfer functions

Note that equation 2.15 predicts an infinite attenuation at the angular frequency ω_0.

A band reject filter can be derived from an inverting band pass filter using a circuit of the form shown in figure 2.14. The circuit action is readily analysed in terms of the usual ideal amplifier assumptions. Thus

$$I_2 + I_1 = I_f$$

$$\frac{V_i}{R_2} - \frac{A_{(s)}}{bp} \frac{V_i}{R_1} = \frac{V_o}{R_3}$$

and

$$\frac{V_{o(s)}}{V_i} = -\frac{R_3}{R_2}\left[1 - \frac{A_{(s)}}{bp}\frac{R_2}{R_1}\right]$$

We substitute for $A_{(s)}$ using equation A.25 and obtain

$$\frac{V_{o(s)}}{V_i} = \frac{R_3}{R_2}\left[\frac{s^2 + \omega_0^2 + b\,\omega_0\,s\left(1 - \frac{A_0}{bp}\frac{R_2}{R_1}\right)}{s^2 + \omega_0^2 + b\,\omega_0\,s}\right]$$

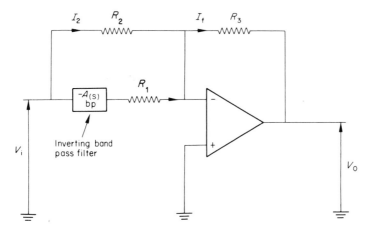

Fig. 2.14 Band reject filter derived from band pass filter

If resistors R_2 and R_1 are proportioned so that $R_1/R_2 = \frac{A_0}{bp}$ the circuit exhibits the required band reject characteristic; the attenuation at the angular frequency ω_0 is determined by how closely this condition is satisfied. Note that the band pass network gives its maximum signal output at the rejection frequency, the output voltage capabilities of the band pass filter thus limits the amplitude of the input signal that can be rejected by the circuit of figure 2.14.

2.6.1 Twin T Band Rejection Filter

The passive RC network shown in figure 2.15 is a so-called 'twin T' or 'parallel – T network', it exhibits a band reject characteristic. The unloaded network introduces no attenuation at zero frequency or infinite frequency and theoretically can be designed to have any desired amount of attenuation at a centre frequency. In practice the amount of attenuation obtainable with the network is determined by component tolerance.

The theoretical condition for infinite attenuation at an angular frequency $\omega_0 = 1/CR$ is that the resistors and capacitors in the circuit should be proportioned as shown in figure 2.15 with

$$a = \frac{k}{k+1}$$

The transfer function for the network can then be written as

$$\frac{V_{o(s)}}{V_i} = \frac{\omega_0^2 + s^2}{s^2 + 2\left[1 + \frac{1}{k}\right]\omega_0 s + \omega_0^2} \qquad (2.17)$$

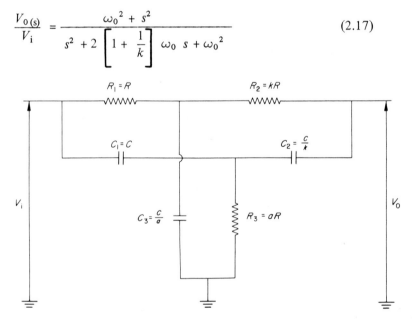

Fig. 2.15 Twin-T passive band reject filter

Note that the transfer function parameter

$$b = 2\left[1 + \frac{1}{k}\right]$$

and the maximum Q value obtainable with the passive circuit is thus

$$Q_{max} = \frac{1}{b_{min}} \simeq \frac{1}{2}$$

A modification to the twin T suggested by Farrer[6], is to connect the junction of R_3 and C_3, which normally returns to ground, to the output voltage or some fraction, m, of the output voltage. The transfer function for the twin T with this modification is

$$\frac{V_{o(s)}}{V_i} = \frac{\omega_0^2 + s^2}{s^2 + (1-m)\,2\left[1 + \frac{1}{k}\right]\omega_0 s + \omega_0^2} \qquad (2.18)$$

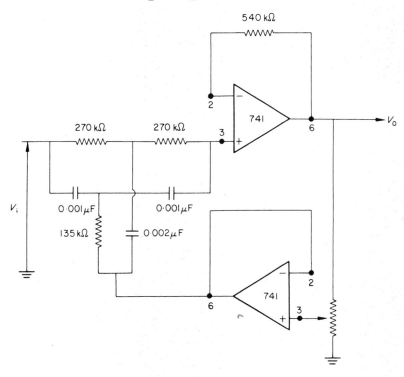

Fig. 2.16 Twin-T band reject active filter

and as the fraction m is made to approach unity the rejection bandwidth becomes extremely narrow. A practical band reject filter (notch filter) which incorporates the modification is shown in figure 2.16. The high input impedance of the first follower-connected operational amplifier prevents loading of the twin T and at the same time permits the use of large resistor values so that comparatively small capacitor values can be used even at low frequencies. The second follower provides a low resistance source for the feedback signal to the twin T and allows a control of the rejection bandwidth (it allows variation of m and hence control of Q).

Design Procedure for Active Twin T Notch Filter

Make $k = 1$ and hence $a = \dfrac{1}{2}$

Choose $C_1 = C_2 = C$ a convenient value

Make $C_3 = \dfrac{C}{a} = 2C$

Make $R_1 = R_2 = R$ calculate the value required from $R = \dfrac{1}{\omega_0 C}$

Make $R_3 = \dfrac{R}{2}$

The circuit of figure 2.16 was connected up for evaluation, resistors were selected to a tolerance of 0.5 per cent and capacitors of 1 per cent tolerance were used. The response of the circuit for $m = 1$ and $m = 0$ is shown in figure 2.17. The closer m is to unity the higher is the Q of the circuit. Varying the potentiometer setting produced responses which lay between the two extremes shown. The depth of the notch which can be obtained is dependent upon component tolerance.

2.6.2 Practical Considerations Governing the Choice of Q

Notch filters are used in order to eliminate unwanted signals with frequencies in a narrow range, ideally they should not seriously affect signals outside this range. Most of the amplitude attenuation of sinusoidal signals takes place for frequencies within the rejection bandwidth but the phase changes produced by a notch are more spread out. Phase shift curves are shown in figure 2.18 (plots of equation 2.16). This suggests that to avoid excessive phase distortion the fundamental frequency of a desired non-sinusoidal signal waveform must lie well outside the rejection bandwidth.

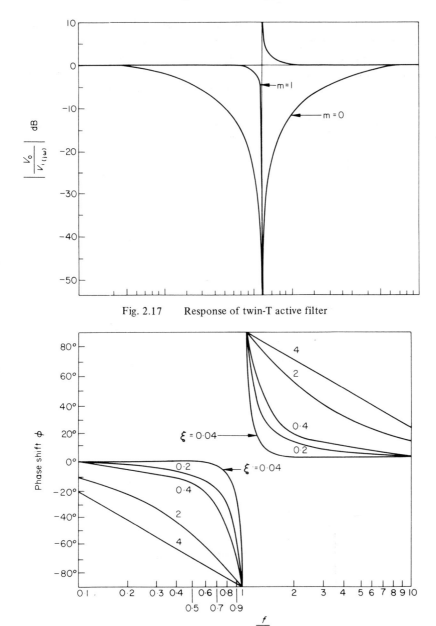

Fig. 2.17   Response of twin-T active filter

Fig. 2.18   Phase shift curves for notch filter

In figure 2.19 we show the effect of the practical notch filter on a triangular waveform. For the upper trace the Q was set at a value of 3.6 and the frequency of the triangular wave was adjusted to be at the lower bandwidth limit. The Q was then changed to a value of 11 with the frequency of the triangular wave held constant, the trace obtained shows considerably less distortion.

The response of a band reject filter to an input voltage step of magnitude E is determined theoretically by the relationship

$$V_{o(t)} = E \left[1 - \frac{b}{\left(1 - \frac{b^2}{4}\right)^{1/2}} e^{-b\pi f_0 t} \sin \beta t \right] \quad (2.19)$$

$$\beta = \pi f_0 \, (4 - b^2)^{1/2}$$

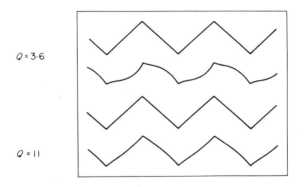

Fig. 2.19   Distortion of triangular wave produced by notch filter

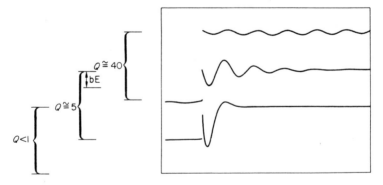

Fig. 2.20   Step response of notch filter for different values of Q

Experimentally observed transient responses are shown in figure 2.20. The traces were obtained for three different settings of Q, the higher the Q, ($Q = 1/b$), the smaller the amplitude of the ringing but the slower the rate of decay. It follows from equation 2.19 that for $b < 1$ the size of the first undershoot in the step response approximates the magnitude $b\,E$ and this provides a convenient method for the experimental determination of Q.

The Q value to be used in a particular filter must be decided with consideration of the conflicting requirements of overshoot and decay time. Another factor which does not favour the use of very high Q rejection filters is their extreme sensitivity to parameter variations. For high Q rejection filters a very small change in the frequency of the signal that it is desired to reject can cause a pronounced change in the attenuation by the filter.

2.7 Multipole Filters – Different Types of Response

In the preceding sections second order filters are the highest order filters treated. The order of a filter is determined by the complexity of the polynomial which describes its characteristic and is equal to the number of poles in the filter transfer function. It is useful to note that the final attenuation rate in the stop band of a high pass or low pass filter is 20 dB per decade, per pole of the filter transfer function. Thus, second order filters give a final rate of attenuation of 40 dB/decade, third order filters give 60 dB/decade, and so on.

A rapid rate of attenuation in the stop band requires the use of a high order filter; high order filter transfer functions can be realised by cascading first and second order filters. An even order transfer function requires a cascade of second order functions, an odd order requires a cascade of a first order function together with second order functions. Consider, for example, high order low pass filters; any even order low pass function can be put in the form

$$A_{(s)} = \prod_{i=1}^{n/2} \left[\frac{A_{oi}}{1 + \dfrac{bi}{\omega_{oi}} s + \dfrac{s^2}{\omega_{oi}^2}} \right] \qquad (2.20)$$

and an odd order in the form.

$$A_{(s)} = \left[\frac{A^1}{1 + \dfrac{s}{\omega_0^1}} \right] \prod_{i=1}^{(n-1)/2} \left[\frac{A_{oi}}{1 + \dfrac{bi}{\omega_{oi}} s + \dfrac{s^2}{\omega_{oi}^2}} \right] \qquad (2.21)$$

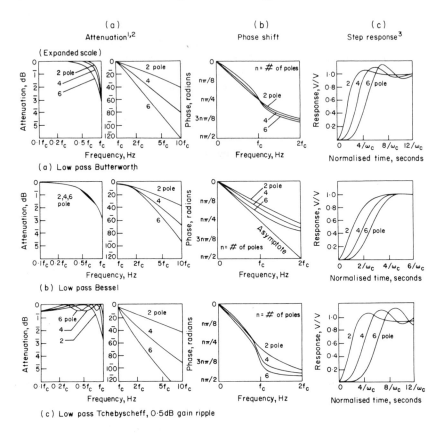

Fig. 2.21 Idealised response curves. General notes; 1. High pass filter responses are obtained as the mirror images of the equivalent low pass responses, 2. Two attenuation scales are used to present greater detail around f_c. 3. Normalised time = $1\omega_c$; $\omega_c = 2\pi f_c$

2.7.1. *Some Filter Designs*

Characteristics other than the final rate of attenuation in the stop band are often of importance in practical filter applications. The ideal filter would pass unchanged all signals with frequencies in the pass band and completely reject signal frequencies in the stop band; real filters do not give such a sharply defined transition between pass and stop bands. Various designs have been devised to optimise specific aspects of the cut-off characteristics of practical filters, the names Butterworth, Tchbeyscheff and Bessel refer to three such designs.

Linear Integrated Circuit Applications

The Butterworth and Tchebyscheff designs are attempts to approximate an ideal magnitude/frequency characteristic. The Butterworth design gives a magnitude/frequency response which is said to be maximally flat in the pass band, at the so-called cut-off frequency, ω_c, separating the pass and stop bands the attenuation is 3 dB. The Tchebyscheff design achieves a magnitude/frequency response with a sharper cut-off characteristic than the Butterworth but at the expense of a pass band which is not completely flat. The Tchebyscheff design gives a ripple in the pass band, the cut-off frequency in this design is often defined as that frequency at which the magnitude response at the pass band edge is down by an amount equal to the ripple in the pass band.

The Butterworth and Tchebyscheff designs are not concerned with signal phase, but phase relationships are important in determining the effect that a filter has on a complex waveform. It is desirable, in order to minimise phase distortion, that the phase shift, ϕ, produced by a filter should vary linearly with frequency. A linear phase variation with frequency gives a group delay $\frac{d\phi}{d\omega}(\omega)$ which is a constant, resulting in equal time delays for each frequency component of a complex wave. In the case of a low pass filter a non-linear phase relationship can give the filter transient response pronounced overshoot and ringing.

The Bessel filter design is an approximation of the ideal linear phase/frequency characteristic but its magnitude/frequency response is a much poorer approximation of the ideal than either the Butterworth or Tchebyscheff designs.

Characteristics of the Butterworth, Bessel and Tchebyscheff filter designs are given in figure 2.21 where for the sake of comparison all curves are plotted with respect to a $-$ 3 dB cut-off frequency f_c.

Practical Design Procedure

The different types of filter response discussed in the previous section represent transfer function polynomials which are mathematically designed to exhibit particular characteristics. Expressions for the response poles are obtained by factoring the polynomials. A knowledge of the poles allows *the bi's and* ω_{0i} *'s* of Eq. 20 and 21 to be determined. It is these parameters which are of importance to the practical designer, their values for the three different types of response discussed, are given in tabulated form. (Table 2.1) In order to illustrate the use of the tables three examples of practical filter designs are given:

Design Example 1. Butterworth Filter. Required, a four pole low pass Butterworth filter with fc = 500 Hz. The design requires a cascade of two second order low pass filters.

From the tables we have

Stage 1
$\omega_{01} = 1\omega_c = 2\pi \times 500$
$b_1 = 1.8478$
$-3\text{dB frequency} = 0.719 f_c = 359.5$ Hz

Use the VCVS circuit of figure 2.2

Make $C_1 = C_2 = C = 0.047$ μF
(Used capacitors of 2 per cent tolerance in practical design)

$R_1 = R_2 = R = \dfrac{1}{\omega_{01} C} = 6.77$ kΩ

In the practical circuit R_1 and R_2 were trimmed by equal amounts to make the response at 5 kHz, (10 f_c), 40 dB down on the response at 50 Hz (0.1 f_c).

$b_1 = 1.8478$ requires a value of $K = 3 - b_1 = 1.15$

The potentiometer was used in order to adjust K to make the response 3 dB down at 360 Hz.

Stage 2
$\omega_{02} = 1\omega_c = 2\pi \times 500$
$b_2 = 0.7654$
peaking frequency $= 0.841 f_c = 420$ Hz
dB of peaking $= 3.01$

The ω_{02} adjustment was performed in the same way as for Stage 1. The potentiometer was adjusted to give 3.01 dB peaking at 420 Hz.
The experimentally obtained response curves are shown in Figure 2.22.

Design Example 2. Tchebyscheff Filters. A four pole low pass filter with a Tchebyscheff response is required. The purpose of these design examples is to illustrate the use of the design tables. In order to simplify experimental procedure an ω_c was designed for which allowed the use of components selected for one of the stages in the previous design example.

From the tables.

Stage 1	*Stage 2*
$\omega_{01} = 0.5286\ \omega_c$	$\omega_{02} = 0.9932\omega_c$
$b_1 = 1.2746$	$b_2 = 0.281$

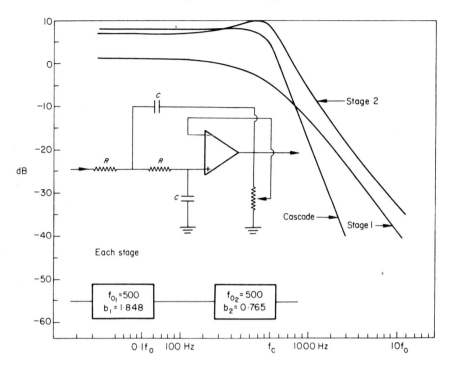

Fig. 2.22   Butterworth design example

peaking frequency = $0.229 f_c$ peaking frequency = $0.973 f_c$
dB of peaking = 0.16 dB   dB of peaking = 11.1 dB

In the practical design illustrated in figure 2.23 f_{o2} was made 500 Hz corresponding to $f_c = \dfrac{500}{0.993}$ = 504 Hz. The f_0 and peaking adjustments were made in the same way as for the previous design example.

Design Example 3. Bessel Filter. A four pole low pass filter with a Bessel response is required. In the Bessel design table the ω_c's are expressed in terms of an amplifier frequency ω_d. ω_d is that frequency for which

$$\frac{d\phi}{d\left(\dfrac{\omega}{\omega_d}\right)} = 1 \text{ second at } \omega = 0.$$ Bessel designs produce a phase shift of 1 radian when $\omega = \omega_d$.

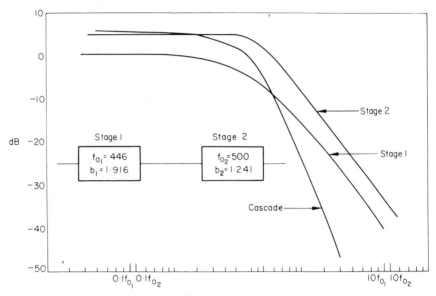

Fig. 2.23   Tchebyscheff design example

From the tables:

Stage 1

$\omega_{01} = 3.0233\ \omega_d$.

$b_1 = 1.9159$

-3 dB frequency $= 2.067\ \omega_d$.

Stage 2

$\omega_{02} = 3.3894\ \omega_d$

$b_2 = 1.2415$

peaking frequency $= 1.624\ \omega_d$.

dB of peaking $= 0.23$ dB

In the practical design example presented in Figure 2.24 f_{02} was made 500 Hz corresponding to $f_d = \dfrac{500}{3.389} = 147.5$ Hz

$f_{01} = 3.023 \times 147.5 = 446$ Hz
1st Stage $-$ 3 dB frequency $= 2.067 \times 147.5 = 305$ Hz
2nd Stage peaking frequency $= 1.624 \times 147.5 = 299.5$ Hz

The f_0 and peaking adjustments were made in the same way as for the previous design examples.

Table 2.1 Filter Design Tables
Butterworth Designs

Number of poles in filter response	Stage	Required value of b_i	Required value of $\dfrac{\omega_{oi}}{\omega_c}$	Peaking frequency or -3dB freq. $\dfrac{\omega_{bi}}{\omega_c}$	dB of peaking $20\log_{10}\dfrac{A(\omega b)}{A(o)}$
2	1	1.4142	1.0000	1.000	
3	1	real pole	1.0000	1.000	
	2	1.0000	1.0000	0.707	1.25
4	1	1.8478	1.0000	0.719	
	2	0.7654	1.0000	0.841	3.01
5	1	real pole	1.0000	1.000	
	2	1.6180	1.0000	0.859	
	3	0.6180	1.0000	0.899	4.62

Tchebyscheff Designs

1 dB ripple

	Stage	b_i	ω_{oi}/ω_c	ω_{bi}/ω_c	dB peaking
2	1	1.0455	1.0500	0.707	1.00
3	1	real pole	0.4942	0.494	
	2	0.4956	0.9971	0.934	6.37
4	1	1.2746	0.5286	0.229	0.16
	2	0.2810	0.9932	0.973	11.1
	1	real pole	0.2895	0.289	
5	2	0.7149	0.6552	0.565	3.51
	3	0.1800	0.9941	0.986	14.93

Bessel Designs

	Stage	b_i	$\dfrac{\omega_{oi}}{\omega d}$	$\dfrac{\omega_{bi}}{\omega d}$	dB peaking
2	1	1.7321	1.7321	1.362	
3	1	real pole	2.3222	2.322	
	2	1.4471	2.5415	2.483	
4	1	1.9159	3.0233	2.067	
	2	1.2415	3.3894	1.624	0.23
	1	real pole	3.6467	3.647	
5	2	1.7745	3.7779	2.874	
	3	1.0911	4.2610	2.711	0.78

See Reference 5 for more comprehensive design tables and higher order filters.

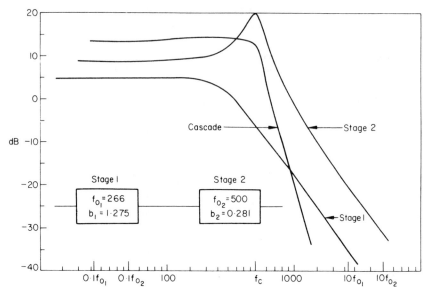

Fig. 2.24 Bessel design example

References

1. L.P. Huelsman, *Theory and Design of Active R.C. Networks*, McGraw-Hill (1968)

2. K.L. Su, *Active Network Synthesis*, McGraw-Hill (1965)

3. S.K. Mitra, *Analysis and Synthesis of Linear Active Networks*, Wiley, New York (1969)

4. G.B. Clayton, *Operational Amplifiers*, Ch. 9 Butterworth, London (1971)

5. Tobey, Graeme and Huelsman, *Operational Amplifiers Design and Applications*, McGraw-Hill (1971)

6. Farrer, *Elect Engng.*, **139** No. 470 p. 219

Exercises 2

2.1 Find the component values to be used in figure 2.1 to give a first order low pass filter with cut off frequency 100 Hz input resistance 50kΩ and gain 10. If the amplifier with the characteristics given in Exercises 1 is used in the circuit what is the input offset error?

2.2 Find the component values to be used for a *VCVS* realisation of a second order low pass filter with the following characteristics,
$A_0 = 10, f_0 = 100$ Hz
$$\zeta = \frac{1}{\sqrt{2}}$$
Use capacitors of value 0.1 μF.

2.3 In the low pass filter circuit given in figure 2.4, $C_1 = 0.01$ μF. $C_2 = 0.005$ μF. Calculate the values to be used for R_1 and R_2 for $f_0 = 10$ Hz, $b = \sqrt{2}$. If the amplifier used in the circuit has the characteristics given in Exercises 1 what is the input offset error.
(a) with no offset balance,
(b) with initial offset balance and a temperature change of 10°C.

2.4 Find the component values to be used in the circuit of figure 2.9 for a band pass transfer function with $f_0 = 100$ Hz, $Q = 5$. What is the gain of the circuit at 100 Hz?

2.5 The following component values are used in the analogue computing loop of figure 2.12; $R = 10$kΩ, $R_1 = 10$kΩ, $R_2 = 100$kΩ, $C = 0.01$ μF. Write down the filter transfer functions which are available in the circuit.

2.6 A passive twin T network (figure 2.15) uses the following component values: $R_1 = R_2 = 100$kΩ, $C_1 = C_2 = 0.01$ μF, $C_3 = 0.02$ μF. Write down the theoretical transfer function for the network.

The network is used with two operational amplifiers in the arrangement shown in figure 2.16 with a fraction $m = 0.95$ of the output voltage fed back. What is the theoretical transfer function of this arrangement? In each case give a dB/log ω plot showing the amplitude response of the circuit.

If the operational amplifiers used in the circuit have the characteristics given in Exercises 1 what is the input offset error with and without a bias current compensating resistor in the circuit?

2.7 The output signal produced by a second order low pass filter is found to exhibit overshoot and ringing in response to a step input signal. With a unit step input the first maximum of 1.53 volts occurs after 3.21 milliseconds, the first minimum of 0.723 volts occurs after 6.41 milliseconds, and the second maximum of 1.15 volts occurs after 9.62 milliseconds. Find the value of the damping factor and the natural frequency for the filter. What will be the small signal settling time of the circuit to 0.1 per cent accuracy. What will be the amount of peaking to be expected in the magnitude response of the filter?

3. Monolithic Timing and Waveform Generator Devices

3.1 Monolithic Timing Devices

Multivibrator circuits are often used in instrumentation systems to perform timing functions, either in generating repetitive timing pulses (free running operation), or in producing single timing periods in response to an input triggering signal (one shot operation). In applications of this kind there are advantages to be gained by using a monolithic integrated circuit which is specifically designed for timing applications. The use of such devices usually involves fewer circuit connections than are necessary with discrete components or operational amplifier multivibrators and can give improvements in performance.

Several monolithic timing circuits are available, the devices generate precise time delays or timing pulses and their timing periods can be varied over a wide range by a choice of externally connected resistance and capacitance values. The devices can be operated in a self-triggering mode to give a free running multivibrator action or they can be externally triggered for a monostable mode of operation. The integrated circuit timer that, currently, has received the most general acceptance is the 555 device which is to be described.

In the 555 timing periods are set by the exponential charging of a capacitor; other devices are available, (e.g. XR 220/320), which include a current source on the device chip for linear capacitor charging. Readers requiring a timing device which produces a linear ramp are advised to consult the manufacturer's data sheet.

3.1.1. *The 555 Timer, Free Running Operation*

The operating principles of the 555 Timer may be understood in terms of the simplified functional schematic of the device given in figure 3.1, external connections for a free running, self-triggering mode of operation are shown.

The device consists essentially of two comparators, a flip-flop, a switch transistor and an output stage. An externally connected timing capacitor is

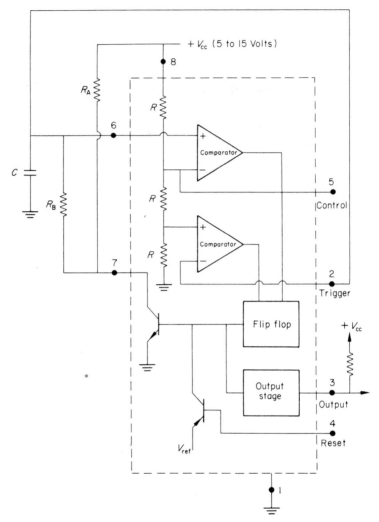

Fig. 3.1   555 timer functional schematic; external connections for free-running operation

charged up towards the positive supply voltage through an external resistor and is discharged by the device switch transistor; the switch transistor is controlled by the output of the flip-flop and the flip-flop itself is set and reset by the output from the comparators. The reference level of one comparator is fixed at a $1/3$ V_{cc} and the other is fixed at $2/3$ V_{cc}, these levels are maintained by three equal resistors in the device which are connected across the supply voltage V_{cc}.

Two sets of waveforms produced by the device for two different values of the supply voltage are shown in figure 3.2, component values are those indicated in figure 3.1. Inspections of the waveforms shows that a change in supply voltage produces a change in waveform amplitude but no observable change in time periods. During the run-up period of the capacitor waveform the switch transistor is held open by the flip-flop and the capacitor charges through the series connected resistors R_A and R_B. When the voltage across the capacitor reaches the reference level of the upper comparator ($2/3$ V_{cc}) the comparator changes the state of the flip-flop and this closes the switch transistor. The capacitor charges down through resistor R_B until its voltage reaches the reference level of the lower comparator ($1/3$ V_{cc}), this comparator changes the state of the flip-flop again, which in turn opens the switch transistor and the cycle repeats.

The charging time is determined by the equation

$$T_1 = C(R_A + R_B) \log_e \frac{V_{cc} - 1/3\ V_{cc}}{V_{cc} - 2/3\ V_{cc}}$$

$$= C(R_A + R_B) \log_e 2$$

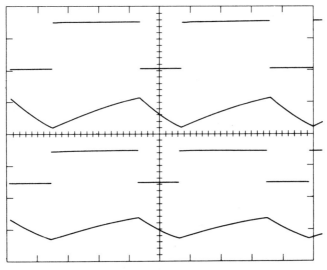

Fig. 3.2   555 timer; waveforms in self-triggering mode of operation. Vertical sensitivities; capacitor waveforms 5V/division; output waveforms 10V/division. Horizontal scale 50µs/division. Upper traces Vcc set at + 15V, Lower traces Vcc set at + 10V

or

$$T_1 \simeq 0.7 \, C \, (R_A + R_B)$$

The output terminal is in its high state during this charging up time. The charging down time period T_2 is determined by the equation

$$T_2 = CR_B \, \log_e \, \frac{0 - 2/3 \, V_{cc}}{0 - 1/3 \, V_{cc}}$$

or

$$T_2 \simeq 0.7 \, C R_B$$

The output is in its low state during the charging down time.

When the 555 is operated in the free running mode oscillations can be switched on and off by a signal applied to the reset terminal. If the reset terminal is grounded the timing capacitor is held in the discharge condition and oscillations are interrupted, oscillations recommence when the reset signal is removed. The reset terminal should be connected to the positive supply line when not in use. If it is required to use the 555 to produce a square wave with unity mark-space ratio the external circuit conditions must ensure equal charge and discharge resistors. Diodes may be used to select charge and discharge paths as shown in figure 3.3 where the charge up path is through diode $D1$ and the charge down path is through diode $D2$.

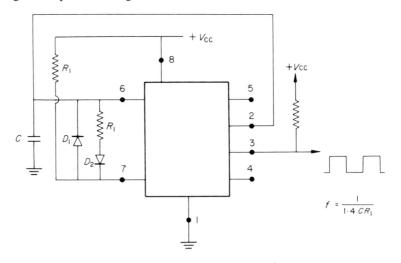

Fig. 3.3   555 timer; external connections for free running operation to give a unity mark-space square wave

The charging up time period and the width of the positive–going part of the square wave output signal can be varied by means of a voltage applied externally to the control terminal. An external control voltage V_{in} in effect overrides the internal reference levels of the two comparators, it sets the reference level of the upper comparator at a value V_{in} and that of the lower comparator at $V_{in}/2$. The charging up time period becomes

$$T_1 = C(R_A + R_B) \log_e \frac{V_{cc} - V_{in}/2}{V_{cc} - V_{in}}$$

Note T_1's dependence on V_{in} is not linear. The charging down period T_2 is now

$$T_2 = CR_B \log_e \frac{0 - V_{in}}{0 - V_{in}/2}$$

or

$$T_2 \simeq 0.7\, CR_B$$

Note that T_2 is not affected by changes in the control voltage. With $V_{cc} = 15$ volts waveforms for two different values (+ 12 volts and + 8 volts) of an externally applied control voltage are shown in figure 3.4. If a varying control voltage is applied the position of the negative going output pulses vary with the magnitude of the control voltage and the device performs the function of a pulse position modulator.

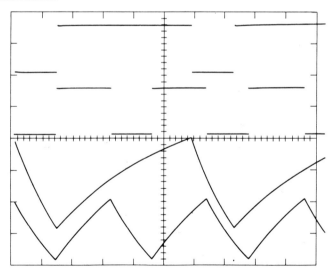

Fig. 3.4 555 timer; effect of control voltage on free running waveforms. Scaling as for figure 3.2; Vcc set at + 15V; V_{in} set at + 12V and + 8V

3.1.2. The 555 Timer, Monostable Operation.

Typical external connections for monostable operation of the 555 device are shown in figure 3.5. The d.c. voltage level at the trigger terminal should be set at a value above the threshold level of the lower comparator ($1/3\ V_{cc}$) and, in the absence of a triggering pulse, the timing capacitor is held in the discharged condition (output low). The circuit triggers on a negative going slope as the voltage at the trigger terminal passes through the value $1/3\ V_{cc}$. Upon triggering the flip-flop is set to remove the short circuit across the capacitor, the capacitor then charges up exponentially through R towards V_{cc} with the time constant CR. When the voltage across the capacitor reaches the threshold level of the upper comparator ($2/3\ V_{cc}$) the flip-flop is reset, the switch transistor is reactivated, the capacitor discharges rapidly towards ground and the cycle is completed. Once the circuit is triggered it is insensitive to further triggering pulses until the timing period is completed, the triggering pulse width must be less than the timing period for proper operation. The timing period may be interrupted by connecting the reset terminal to ground, this turns on the switch transistor which prevents the capacitor from charging.

The duration of the timing period T during which the output level is at a high state is determined by the relationship

$$T = CR \log_e \frac{V_{cc} - 0}{V_{cc} - 2/3\ V_{cc}}$$

or

$$T \simeq 1.1\ CR$$

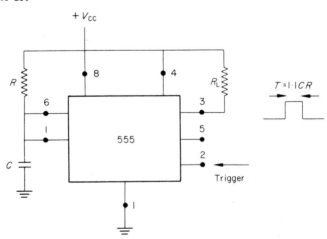

Fig. 3.5   555 timer; external connections for monostable operation

Linear Integrated Circuit Applications 97

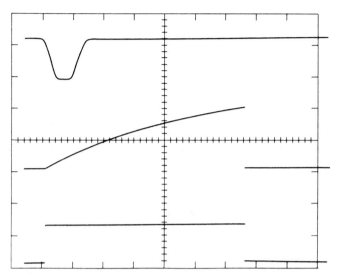

Fig. 3.6   555 timer; monostable waveforms. Upper trace trigger (pin 2) 1V/division; middle trace capacitor waveform (pin 6, 9 and 7) 2V/division; lower trace output (pin 3) 5V/division. Horizontal scale 0.2 ms/division. C = 0.01 μF, R_A = 100kΩ, V_{cc} set at + 6V

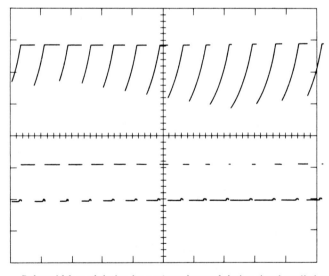

Fig. 3.7   Pulse width modulation by a triangular modulating signal applied to pin 5 via a 10kΩ resistor

98 Linear Integrated Circuit Applications

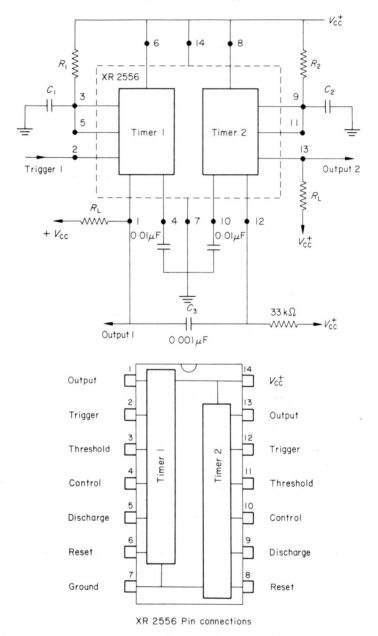

Fig. 3.8   Sequential timing with dual 555 timer

Monostable waveforms obtained with the circuit of figure 3.5 using $C = 0.01 \mu F$. $R = 100k\Omega$ and $V_{cc} = 6$ volts are shown in figure 3.6.

The monostable timing period can be changed (but not linearly) by a voltage applied to the control terminal. If the 555, connected in the monostable mode, is triggered by a continuous pulse train, the duration of the timing cycle (i.e. the output pulse width) can be modulated by applying a modulation input to the control terminal. Waveforms illustrating a modulation of pulse width are shown in figure 3.7.

An increased versatility in switching and timing applications may be obtained by combining two 555 timers. It is convenient to use a dual 555 for this type of application; the XR 2556 is such a device, it contains two independent 555's on a single monolithic chip and is available in a 14 pin dual in-line package.

3.1.3. Sequential Timing

The output of one timer section may be used to trigger a second timer for a sequential timing operation, the circuit connections to be made to the XR 2256 are shown in figure 3.8. When timer 1 is triggered at pin 2, its output at

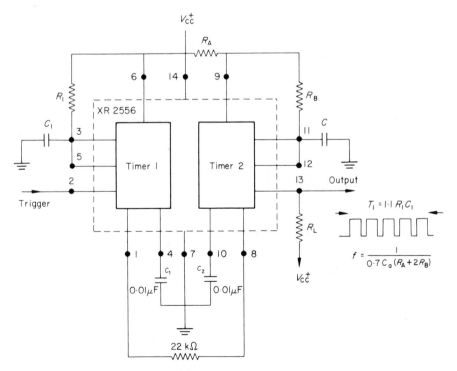

Fig. 3.9　Keyed oscillator using dual 555 timer

pin 1 goes high for a time duration $T_1 = 1.1\, R_1\, C_1$. At the end of this timing cycle, pin 1 goes low and triggers timer 2 through the capacitive coupling C_3 between pins 1 and 12. The output at pin 13 then goes high for a time duration $T_2 = 1.1\, R_2\, C_2$ and in this way the unit behaves as a 'delayed one shot' where the output of timer 2 is delayed from the initial trigger at pin 2 by a time delay T_1.

3.1.4. Keyed Oscillator

One timer operated as a monostable can be used to switch on and off a second timer operating in its free running mode, circuit connections are shown in figure 3.9. Timer 1 operates as a monostable, its output is connected to the reset terminal of timer 2. Under rest conditions the output of timer 1 is low and operation of timer 2 is inhibited. Upon the application of a triggering signal to timer 1 it's output goes high and the oscillator section, timer 2, is switched on. The output of timer 2 appears as a tone burst with frequency determined by R_A R_B and C_2 and duration set by R_1 and C_1.

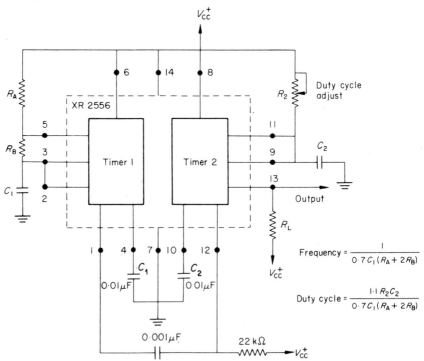

Fig. 3.10   Fixed frequency, variable duty cycle oscillator with dual 555 timer

3.1.5. *Fixed Frequency – Variable Duty Cycle Oscillator*

The circuit connections shown in figure 3.10 allow a dual 555 to be used as a fixed frequency variable duty cycle square wave generator. The first timing section, operating in a free running mode, is used to continuously trigger the second timer operating in a monostable mode. The timing period of the monostable T_2 is made less than the period of the oscillations produced by timer 1. The output of timer 2 is thus a square wave with the same frequency as that produced by timer 1 but with a duty cycle that can be varied over a wide range, (1 per cent to 99 per cent) by adjustment of R_2.

3.2 Monolithic Waveform Generators.

The basic wave shapes produced by these devices are triangular and square, they are generated as a result of the linear charging and discharging of a capacitor with comparators, or some form of trigger circuit, used to switch between the charge and discharge conditions. Frequency determining components are externally connected, the devices include provision for a control of frequency by means of an externally applied control voltage thus allowing for frequency sweep and frequency modulation. Some devices include internal circuitry used to shape the triangular waves to sine waves, in others, this operation must be performed externally. Specific devices and the practical considerations which are involved in their use will be considered.

3.2.1. *The 566 Waveform Generator (Signetics SE 566 T, NE 566 T, NE 566 V)*

The 566 device is a general purpose VCO designed for linear frequency modulation; it provides simultaneous triangular and square wave outputs at frequencies up to 1 MHz. If a sinusoidal waveform is required the triangular wave may be shaped using an external circuit. A simplified functional schematic of the 566 device is shown in figure 3.11. An external capacitor C_1 connected between pin 7 and ground is charged and discharged by a current source I of a magnitude determined by an external resistor R_1 and a control voltage V_c applied to pin 5. The value of the charging current is determined by the relationship

$$I = \frac{V_s - V_c}{R_1}$$

Values of V_c must be in the range

$$3/4\ V_s \leqslant V_c \leqslant V_s$$

and values of R_1 in the range

$$2k\Omega \leqslant R_1 \leqslant 20k\Omega$$

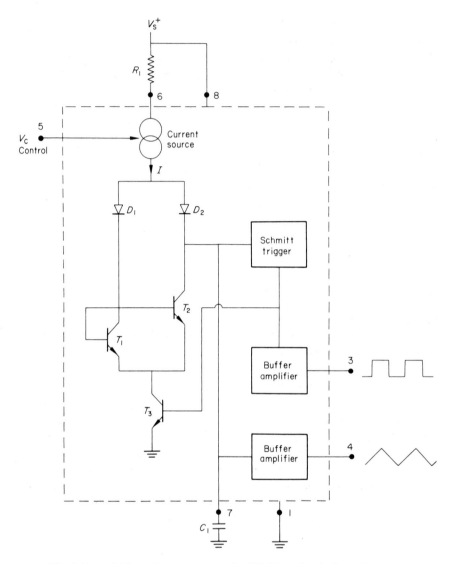

Fig. 3.11   566 waveform generator; simplified functional schematic

Referring to figure 3.11, transistor T_3 is initially off and the current I charges C_1 through diode $D2$. When the voltage across C_1 reaches the upper triggering limit of the Schmitt trigger circuit, the Schmitt circuit changes states and switches transistor T_2 on, this effectively grounds the emitters of T_1 and T_2. The current I now flows through D_1 T_1 and T_3 to ground, but since the base emitter voltages of T_1 and T_2 are the same potential an equal current must flow through T_2. Since D_2 is now reverse biased this current discharges C_1 until the lower trigger level of the Schmitt circuit is reached at which point the cycle repeats.

Capacitor charge and discharge rates are the same, I/C_1 volts per second, and charge and discharge times are determined by the relationship

$$t = \frac{V_H C_1}{I} \quad \text{seconds}$$

V_H is the voltage differential between the upper and lower input transition levels of the Schmitt trigger circuit. The design of the trigger circuit is such that V_H is directly proportional to the supply voltage

$$V_H \simeq \frac{V_s}{5}$$

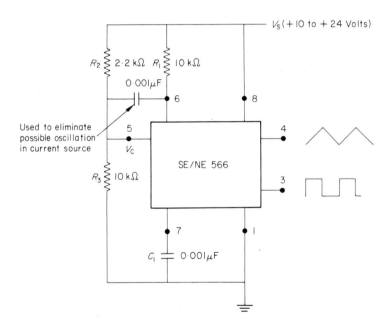

Fig. 3.12   Circuit for fixed frequency operation of the 566

Thus

$$t = \frac{V_s}{5(V_s - V_c)} C_1 R_1$$

and

$$f = \frac{1}{2t} \simeq \frac{5}{2} \frac{V_s - V_c}{V_s C_1 R_1}$$

If the value of V_c is fixed by means of a potential divider connected between the supply line and ground, V_c becomes a fixed fraction of V_s and the frequency of oscillations is then largely independent of the supply voltage. Frequency proportioning over a 10:1 range is allowed by selection of R_1. Frequency can be swept over a 10:1 range with an adjustment of V_c within its allowed range.

A typical circuit arrangement of the 566 is shown in figure 3.12 with waveforms given in figure 3.13. The amplitude of the output waveforms is dependent upon the supply voltage used. The circuit will operate with supply voltages in the range 10 to 24 volts. Some deterioration in waveforms occurs at frequencies approaching the operating limit. Waveforms produced by the circuit of figure 3.12 with the value of C_1 changed to 100pF and R_1 to 2.7 kΩ are shown in figure 3.14.

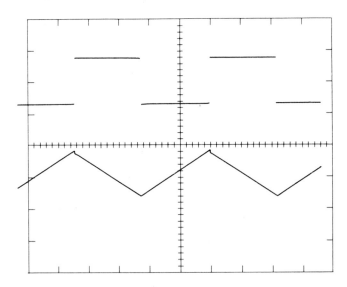

Fig. 3.13 566 waveforms. Square wave 5V/division, scale zero central graticule line; triangular wave 2V/division, scale zero lower graticule line; time scale 5μs/division

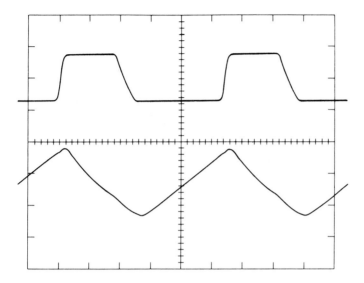

Fig. 3.14   566 waveforms at operating frequency limit; vertical scaling as for figure 3.13; horizontal scaling 0.2 μs/division

3.2.2 Single Cycle and Gated Operation of the 566

The 566 can be used to produce a single wave or group of waves if its normal action is interrupted by applying an external hold condition to the timing capacitor. A circuit modification for triggered operation of the 566 is shown in figure 3.15.

The operational amplifier in the circuit is connected as a bistable, under rest conditions its output is at its negative saturation level. The voltage across the timing capacitor is prevented from rising above a level determined by the setting of potentiometer P_1 and is held at that level. Upon the application of an external triggering pulse to the bistable it switches to its positive saturation state, the hold is removed and the timing capacitor starts to charge. The bistable is reset to its negative saturation state by the negative going part of the 566 output square wave and the timing capacitor stops charging again when its voltage reaches the level set by potentiometer P_1. The 566 produces a single cycle with starting point determined by adjustment of P_1. Waveforms for three different settings of the potentiometer are shown in figure 3.16, the circuit was triggered repetitively by an external square wave.

106 Linear Integrated Circuit Applications

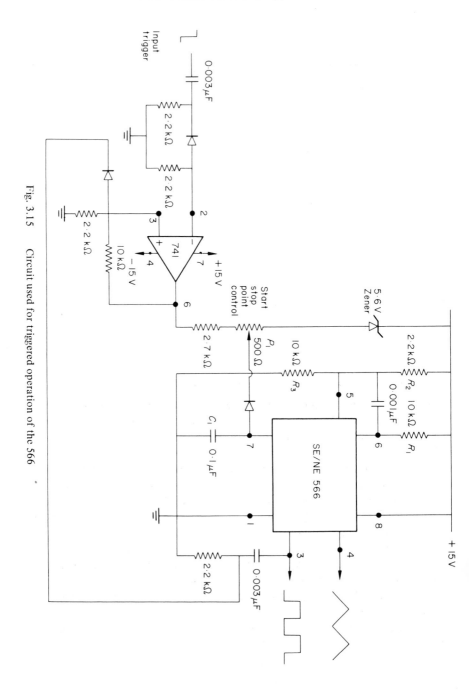

Fig. 3.15 Circuit used for triggered operation of the 566

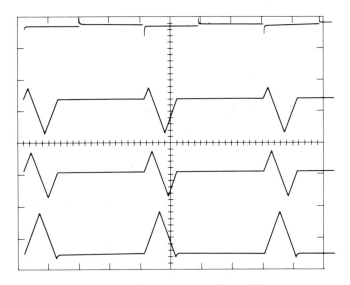

Fig. 3.16 Repetitive triggering of 566; waveforms with different settings of start/stop control

Gated operation of the circuit can be arranged by causing the bistable to remain in its positive saturation state for a preset period, during which period the 566 operates continuously.

3.2.3 The 566 Used as a Ramp Generator

An external transistor which is used rapidly to discharge or charge the timing capacitor can be used to transform the triangular output wave produced by the 566 into a positive — or negative-going ramp. The circuit modification for a positive-going ramp is illustrated in figure 3.17, waveforms produced are shown in figure 3.18. Note that the squarewave output of the 566 is changed to a positive-going pulse. A circuit modification which may be used to change the 566 output waveforms into a negative-going ramp and a negative pulse is shown in figure 3.19.

3.2.4 *Asymmetrical Waveforms with the 566*

The 566 waveforms may be made non-symmetrical by incorporating an external circuit modification which causes capacitor charge and discharge currents to be

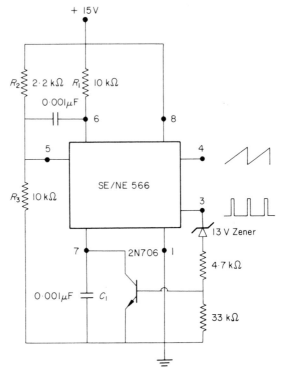

Fig. 3.17   566 used for positive ramp generation

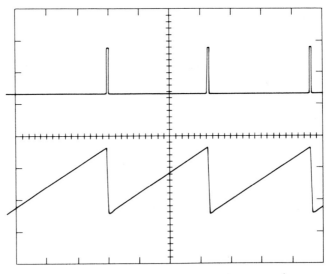

Fig. 3.18   566 ramp waveform; vertical scaling as for figure 3.13, horizontal scaling 5µs/division

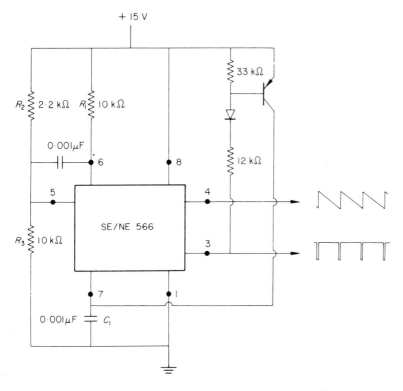

Fig. 3.19   The 566 used for negative ramp generation

different. A circuit which can be used for this purpose is shown in figure 3.20. The external transistor is switched into saturation on the negative-going parts of the 566 square wave output, this effectively reduces the value of the resistor R_1 which determines the capacitor charging current and the capacitor charges up more rapidly than it charges down.

3.2.5 F.M. Generation with two 566's

Two 566 waveform generators may combine for use as an F.M. generator. The triangular output wave produced by one generator is applied to the control terminal of the second generator so as to modulate its frequency, the circuit arrangement is shown in figure 3.21. Capacitor C_1 selects the modulation frequency adjustment range and C_1^1 selects the centre frequency. The value of capacitor C_2 should be chosen so that it has negligible reactance at the

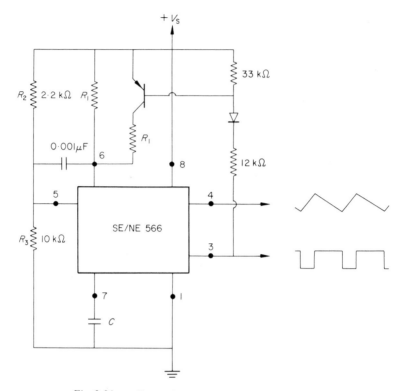

Fig. 3.20 Generation of asymmetrical waveforms

modulation frequency. If a frequency sweep in only one direction is required, the first 566 generator may be modified as described in section 3.2.3 so that it produces a voltage ramp to be applied as a control signal to the second generator.

3.2.6 The Intersill 8038 Waveform Generator

The Intersill 8038 Waveform Generator is another example of an integrated circuit VCO; it is in some respects similar to the 566 device but provides an additional sinuosoidal output waveform produced by internal sine shaping circuitry.

A simplified functional schematic of the 8038 devices is shown in figure 3.22. The device contains two current sources I_1, and I_2 of value set by external resistors R_1 and R_2 and by a control voltage applied to terminal 8. Current source I_1 is supplied as a continuous charging current to an external timing

Linear Integrated Circuit Applications

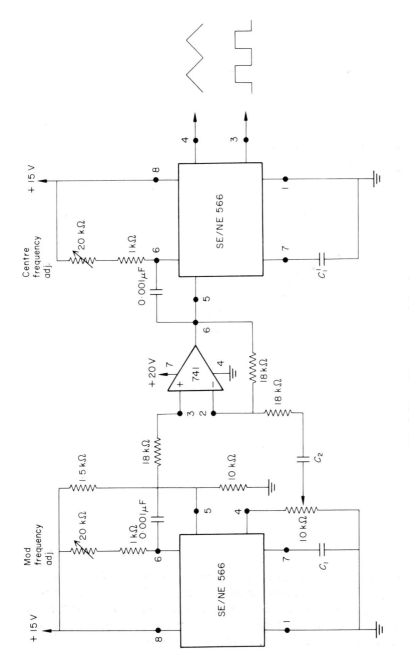

Fig. 3.21 Frequency modulated generator

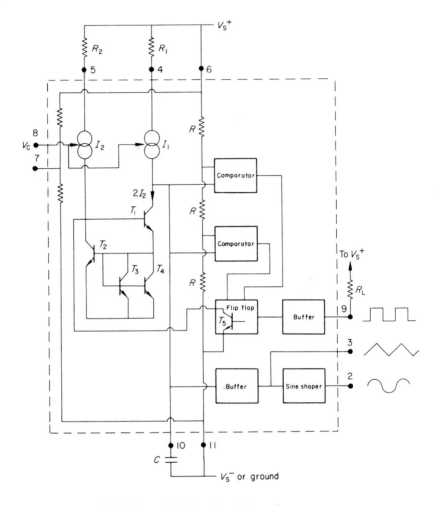

Fig. 3.22   8038 functional schematic

capacitor C. Current source I_2 is converted into a current $2I_2$ by transistors T_1, T_2 and diode connected transistors T_3 and T_4. The current $2I_2$, is switched in as a capacitor discharge current when transistor T_5 of the device flip-flop is switched off. The net discharge current is then $2I_2 - I_1$ and if I_1 and I_2 are made the same, charge and discharge currents are equal.

The flip-flop is set and reset, (T_5 switched on and off), by two comparators used to sense the level of the voltage across the timing capacitor; the threshold levels of the two comparators are set at $2/3\ V_s$ and $1/3\ V_s$ by three equal resistors connected across the supply voltage. Currents I_1 and I_2 are determined by the relationships

$$I_1 = \frac{V_s - V_c}{R_1}$$

$$I_2 = \frac{V_s - V_c}{R_2}$$

The capacitor is charged and discharged between the levels $1/3\ V_s$ and $2/3\ V_s$ giving a triangular wave of magnitude $1/3\ V_s$. During the rising portion of the triangular wave the capacitor is charged by I_1 alone and the rising part of the wave takes place in a time period

$$t_1 = C \frac{V_s}{I_1\ 3}$$

During the falling part of the triangular wave the net discharge current is $2I_2 - I_1$ and the falling part of the triangular wave takes place in a time period

$$t_2 = \frac{C\ V_s}{(2I_2 - I_1)3}$$

If $R_1 = R_2$ then I_1 and I_2 are equal and charge and discharge times are the same, giving a 50 per cent duty cycle and a frequency

$$f = \frac{1}{t_1 + t_2} = \frac{3}{2} \frac{V_s - V_c}{CR_1\ V_s}$$

If pins 7 and 8 are connected together V_c is set at a value $V_c = 4/5\ V_s$ by an internal potential divider making the frequency independent of supply voltage with a value

$$f = \frac{0.3}{CR_1}$$

The 8038 may be operated from a single supply (+ 10 volts to + 30 volts) or equal supplies (± 5 to ± 15 volts); with dual supply operation output waveforms take place about ground potential. Circuit connections for fixed frequency operation are shown in figure 3.23 and the waveforms obtained with this arrangement are shown in figure 3.24. Resistor R_2 was adjusted for a 50 per cent duty cycle, the upper sine wave was obtained with a fixed 82kΩ resistor connected between pin 12 and the negative supply. The lower sine wave was obtained after performing the recommended trim procedure for minimising sine

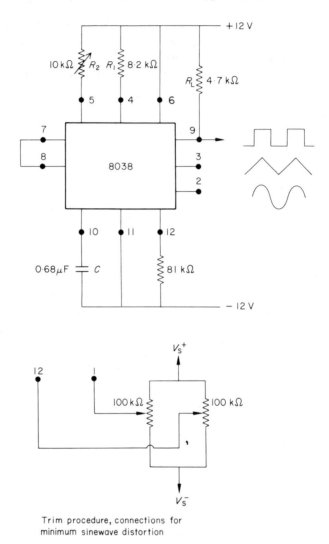

Trim procedure, connections for minimum sinewave distortion

Fig. 3.23   Fixed frequency operation of the 8038

wave distortion, the connections for this adjustment are shown in figure 3.23. Note that the sine wave output from the device is not buffered internally. If the sine wave output is loaded, waveform distortion results making the use of an external buffer circuit necessary. Triangular and sine wave forms show a significant deterioration if component values are changed so as to give an operating frequency approaching the performance limits of the device (1MHz).

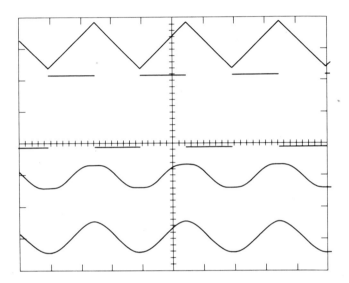

Fig. 3.24 8038 waveforms; square wave at 10V/division, triangular and sinewaves at 5V/division, time scale 5ms/division. Sinewave shown with and without recommended trim procedure

The square wave output (pin 9) is available at the uncommitted collector of an internal transistor and the output load resistor connected to pin 9 may, if required, be returned to a separate power supply. In this way the square wave, for example, can be made TTL compatible with the load resistance returned to + 5 volts whilst the waveform generator is powered from a higher voltage.

Non-symmetrical waveforms may be obtained with the 8038 simply by using different values for resistors R_1 and R_2. The waveforms given in figure 3.25 show the effect of using a value of R_2 significantly less than R_1. Triggered or gated operation of the 8038 can be obtained by applying a hold condition to the capacitor as discussed in section 3.2.2.

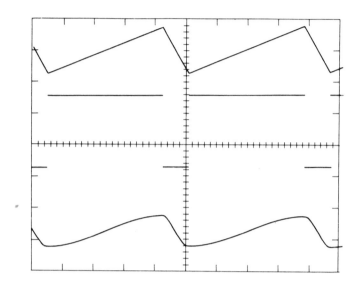

Fig. 3.25   8038 waveforms at $R_2 < R_1$.. Scaling as for figure 3.24

3.2.7 *Frequency Modulation and Frequency Sweeping of the 8038*

The frequency of the waveforms produced by the 8038 is a direct function of the d.c. voltage at terminal 8; by altering this voltage frequency modulation is achieved. For small deviations (e.g. ± 10 per cent) the modulating signal can be applied directly to pin 8 using a capacitor to provide d.c. decoupling. The connections are shown in figure 3.26. The external resistor between pins 7 and 8 is not necessary but it can be used to increase input impedance. Without it the input impedance is 8kΩ, with it this impedance increases to R + 8kΩ.

In applications requiring a large F.M. deviation or a frequency sweep the modulating signal is applied between the positive supply line and pin 8. In this way the entire bias for the current sources is created by the modulating signal and a very large (e.g. 1000 : 1) sweep range is created. Care must be taken to regulate the supply voltage in this configuration for the charge current is no longer a function of the supply voltage (yet the trigger thresholds still are) and thus the frequency becomes dependent upon the supply voltage. The potential on pin 8 may be swept from V_s to about 2/3 V_s. Provision for an earth-referred sweep or modulating voltage can be arranged using the operational amplifier circuit shown in figure 3.27.

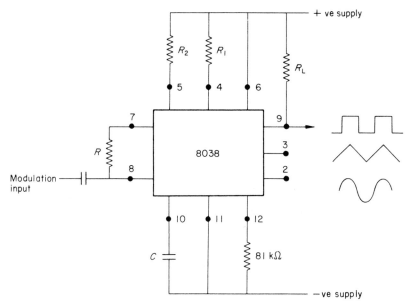

Fig. 3.26   Connections for frequency modulation

Exercises 3.

3.1   The 555 timer is used in the free-running mode of operation with $R_A = 10k\Omega$ and $R_B = 20k\Omega$, $C = 0.01\mu F$ and a +15 volt supply. Calculate the timing period T_1 and T_2 and sketch the waveforms which you would expect the device to give.

3.2   The following component values are used in the circuit of figure 3.10. $R_A = 10k\Omega, R_B = 10k\Omega, C_1 = 0.01~\mu F, C_2 = 0.005~\mu F$. What is the frequency of the oscillations produced? What value of R_2 gives a 20 per cent duty cycle?

3.3   The 566 waveform generator show in figure 3.12 is used with the following component values; $C_1 = 0.001~\mu F, R_1 = 10k\Omega, R_2 = 2k\Omega$, $R_3 = 10k\Omega$, $V_s = 12$ volts. What is the approximate value of the frequency of oscillations?

3.4   The 8038 waveform generator in figure 3.23 is used with the following component values; $R_1 = 5k\Omega, R_2 = 6k\Omega, C = 0.02~\mu F, V_s = 15$ volts. Pins 7 and 8 are connected together.
What are the time periods of the rising (t_1) and falling (t_2) parts of the triangular waveform? Sketch the waveforms to be expected.

118 *Linear Integrated Circuit Applications*

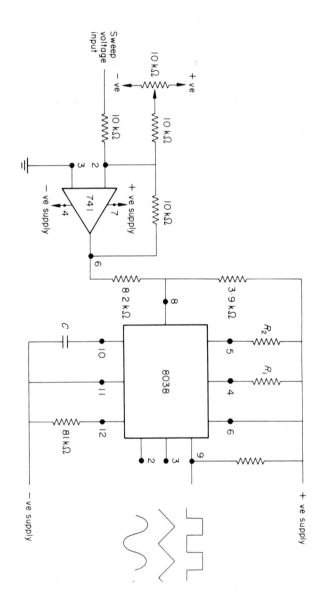

Fig. 3.27 Earth referred frequency sweep

4. Variable Transconductance Devices

The basic monolithic integrated circuit operational amplifier is no doubt the most universally applicable linear i.c. but there are now quite a wide range of other linear integrated circuits available. A general appreciation of the internal circuit operating principles of these devices enables them to be used with greater confidence and leads to an understanding of the function of any external components that are required for their correct operation. Several linear i.c.'s make use of a variable transconductance circuit technique, and this chapter is devoted to a brief theoretical treatment of the variable transconductance principle and discusses the internal circuitry of some devices which use the technique[1]. The technique is basically employed to achieve a variable gain, the variable gain facility is developed to provide a device which can perform a four quadrant linear multiplication.

4.1 The Variable Transconductance of a Bipolar Transistor

Variable transconductance is an inherent property of a bipolar transistor but the effective exploitation of the property requires the use of several transistors whose characteristics must match and track thermally. It is difficult to achieve this matching in circuits made up from discrete components but a close matching of characteristics is a particular feature of the transistors formed in the monolithic integrated circuit manufacturing process.

The transconductance of an amplifying device relates the output current of the device to the input voltage. In the case of a bipolar transistor, used in the common emitter configuration, the output current is the collector current and the base emitter voltage is the input voltage. For small changes in current and voltage, ΔI_C and ΔV_{BE}, the transconductance of the transistor may be written as $\Delta I_C / \Delta V_{BE}$. The collector current of a transistor used with a reverse biased collector base junction can be represented approximately by the equation

$$I_C \cong I_0 \left(\exp \frac{-q V_{EB}}{kT} - 1 \right) \tag{4.1}$$

q is the electronic charge
T is the temperature in K
k is Boltzmans constant.

We write $V_{BE} = -V_{EB}$ and $V_T = \dfrac{kT}{q}$. If values of constants are substituted it is found that $V_T \cong 26$ mV at $27°C$ so that for values of V_{BE} greater than say 100 mV the exponential term in the equation predominates and we may then write the equation as

$$I_C \cong I_O \exp \dfrac{V_{BE}}{V_T} \tag{4.2}$$

Differentiation of equation 4.2 allows us to write the transconductance of the transistor as

$$\dfrac{\Delta I_C}{\Delta V_{BE}} \cong \dfrac{1}{V_T} I_C$$

If we further assume that $I_C \triangleq I_E$ we may write

$$\dfrac{\Delta I_C}{\Delta V_{BE}} \cong \dfrac{1}{V_T} \cdot I_E \tag{4.3}$$

The transconductance and hence the small signal voltage gain of a bipolar transistor is seen to vary linearly with the magnitude of the emitter current, but note the temperature dependence, through the term V_T and a more marked dependence through the I_O term. I_O and hence I_C and I_E approximately double for a $10°C$ rise in temperature if the transistor is not temperature compensated.

4.2 Using Variable Transconductance for Gain Control

Linear i.c.'s with an automatic gain control capability and multiplier/modulator devices exploit the variable transconductance property of bipolar transistors by using a differential amplifier circuit configuration which serves to balance out the effects of the temperature dependence of transistor parameters. The basic action of such devices is analysed in terms of the circuit model illustrated in figure 4.1. The transistors in the circuit are assumed to be identical, transistor base currents are assumed to be negligible in comparison with collector currents so that collector and emitter currents can be assumed equal.

Using equation 4.2 we write

$$I_1 \cong I_0 \exp \dfrac{V_{BE}}{V_T} \quad \text{and} \quad I_2 \cong I_0 \exp \dfrac{V_{BE2}}{V_T}$$

The assumption of negligible base currents allows us to write

$$I_3 = I_1 + I_2$$

and substitution gives

$$I_3 = I_0 \left(\exp \dfrac{V_{BE1}}{V_T} + \exp \dfrac{V_{BE2}}{V_T} \right.$$

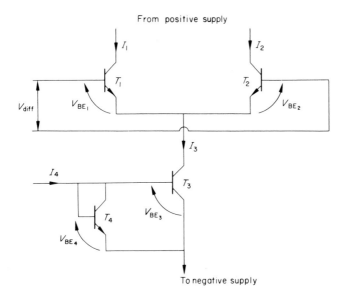

Fig. 4.1

We write $V_{\text{diff}} = V_{BE_1} - V_{BE_2}$ and by rearrangement obtain

$$I_3 = I_0 \exp \frac{V_{BE_2}}{V_T} \left(\exp \frac{V_{\text{diff}}}{V_T} + 1\right)$$

and therefore

$$I_2 = \frac{I_3}{\exp \dfrac{V_{\text{diff}}}{V_T} + 1} \qquad (4.4)$$

Similarly

$$I_1 = \frac{I_3}{\exp -\dfrac{V_{\text{diff}}}{V_T} + 1} \qquad (4.5)$$

Equations 4.4 and 4.5 are used to calculate values of I_1 and I_2 and these values are shown plotted as a function of the differential input voltage in figure 4.2. For convenience V_{diff} is plotted in units of V_T. We see that the current I_3 divides between transistors T_1 and T_2 in proportions governed by V_{diff}. For

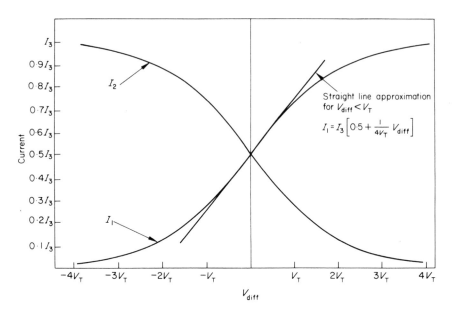

Fig. 4.2

small values of V_{diff}, I_1, I_2 and hence the differential current, $I_1 - I_2$, all vary approximately linearly with V_{diff}. The values of I_1, I_2 and $I_1 - I_2$ for values of V_{diff} less than V_T are determined by the approximate relationships

$$I_1 \cong I_3 \left(0.5 + \frac{1}{4 V_T} V_{diff}\right) \tag{4.6}$$

$$I_2 \cong I_3 \left(0.5 - \frac{1}{4 V_T} V_{diff}\right) \tag{4.7}$$

$$I_1 - I_2 \cong \frac{1}{2 V_T} I_3 V_{diff} \tag{4.8}$$

The differential output current is seen to depend upon the product $I_3 V_{diff}$ and the circuit can be used to perform a multiplication operation, but with definite limitations. Operation is approximately linear only for values of V_{diff} less than V_T; only two quadrant multiplication can be performed for the circuit will not function with the current I_3 reversed.

In figure 4.1 the current I_3 is stabilised against the effect of temperature variations of transistor T_3 by the diode connected transistor T_4. Using equation

Linear Integrated Circuit Applications

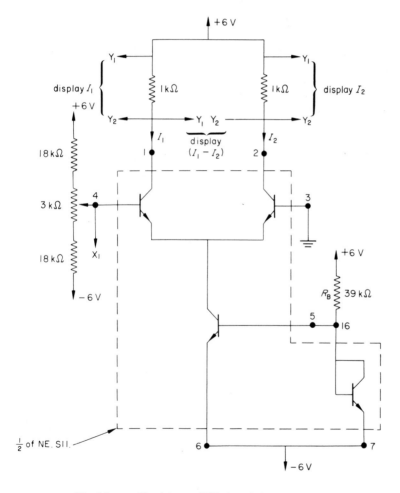

Fig. 4.3 Obtaining an X/Y plot of circuit response

4.2 we may write,

$$I_3 = I_0 \exp \frac{V_{BE_3}}{V_T} ; \quad I_4 = I_0 \exp \frac{V_{BE_4}}{V_T}$$

But
$$V_{BE_3} = V_{BE_4}$$
thus
$$I_3 = I_4$$

The two transistors are assumed to be at the same temperature. A constant value for I_3 is ensured by supplying transistor T_4 with a constant collector current.

A practical test circuit which may be used experimentally to investigate the validity of the above analysis is shown in figure 4.3, the circuit uses one half of the dual differential amplifier device type NE 511 (Signetics). Variations of collector current with change in differential input voltage are displayed as X/Y plots, typical experimentally obtained graphs are given in figures 4.4 and 4.5,

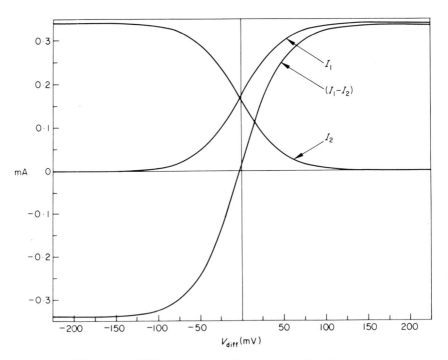

Fig. 4.4 X/Y plots obtained with test circuit; $R_B = 39k\Omega$

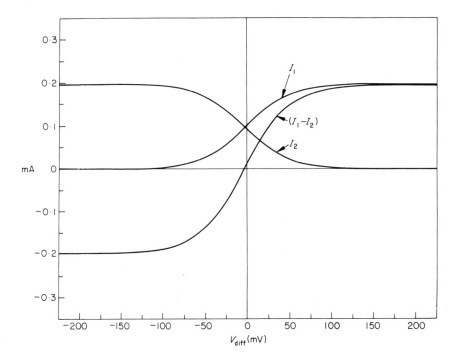

Fig. 4.5   X/Y plots with $R_B = 68k\Omega$

the two sets of curves were obtained for two different values of the resistor R_B, 39kΩ and 68kΩ. R_B sets a value for the bias current I_B and hence the current I_3. The current I_B is related to the value of R_B by the approximate relationship

$$I_B \cong \frac{V_s^+ + |V_s^-| - V_D}{R_B}$$

$V_D \cong 0.75$ V at 25°C

In addition to controlling the differential gain, the bias current I_B also controls the magnitudes of the base bias currents drawn by the differential input terminals of the amplifier.

4.3 Controlled Gain I.C. Devices

The NE 511 device used in the circuit of figure 4.3 has only active components formed on the device chip, the external components which are connected to the device make it function as a differential amplifier. Control of the bias current I_B provides a gain control facility. Alternative arrangements of external passive components allow operation as a common emitter-common base amplifier (cascode configuration), or common collector-common base amplifier all circuits having an AGC capability. The device is intended for RF/IF amplifier applications at frequencies up to 100 MHz. There are several devices which are similar in function to the 511 but which include passive components formed on the device chip, simplifying the external connections required for specific applications. Examples of such devices are; Signetics 510, Motorola MC 1550, RCA CA 3028. The internal circuit details of these devices may be compared by consulting the appropriate manufacturers data sheets.

4.3.1 *Controlled Operational Amplifiers, (Programmable Op. Amps)*

The principle of controlling operating currents by means of an externally supplied bias current is applied in several operational amplifier types. In devices of this type the quiescent operating currents in the several amplifying stages of the device are all controlled by means of an externally supplied bias current, the value of the bias current is normally set by a single external resistor. The arrangement offers the user considerable flexibility in that, by controlling quiescent operating currents, he not only controls the open loop gain of the amplifier but also sets values for other important amplifier parameters such as, power consumption, slew rate, input resistance, input bias current and input offset current. Examples of this type of operational amplifier are; RCA Micropower Op. Amp. type CA 3078, National Programmable Op. Amp. type LM 4250, Fairchild Programmable Op. Amp. type μA 776, and Silkonix Triple Op. Amp. Type L144. Comparison of the detailed internal circuit schematics of the devices is best achieved by reference to the device date sheets.

Another example of a device which uses an external bias setting technique is the RCA, CA 3060. This device consists of an array of three identical independent amplifiers. The amplifiers are referred to as 'Operational Transconductance Amplifiers' since their gain characteristic is best described by transconductance (output current/input voltage); the output circuits of the amplifiers act essentially as current sources. Open loop voltage gain of an amplifier is determined by the product of the transconductance and the load resistance, other characteristics of the amplifiers are similar to those of an operational voltage amplifier but the amplifier parameter magnitudes are a function of the amplifier bias current, I_{ABC}. A simplified circuit schematic of a single amplifier and a functional block diagram of the complete device are shown in figure 4.6.

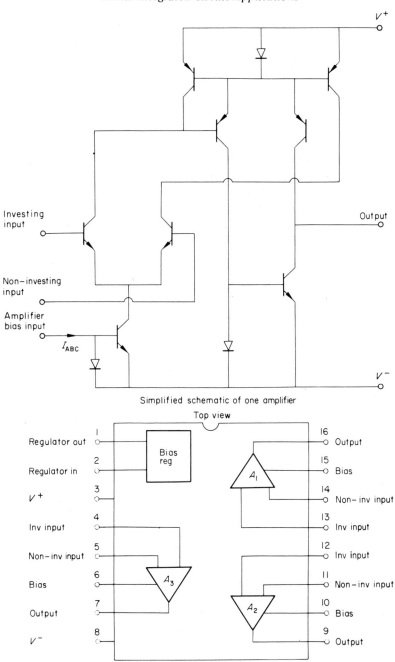

Fig. 4.6 Operational transconductance amplifier array type CA3060

128   *Linear Integrated Cirucit Applications*

Note that controlled operational amplifiers are particularly suitable for use with low voltage power supplies and in battery operated equipment where minimum power consumption is of prime importance.

4.3.2 *The Gate Controlled, Two Channel, Wide-Band Amplifier type MC 1545 (Motorola)*

The MC 1545 device is another example of a linear i.c. whose operating principle is derived essentially from the properties of the basic differential amplifier circuit of figure 4.1. The device has two differential input channels a differential output channel and a single-ended control or gating channel; its circuit schematic is given in figure 4.7. The voltage gain for an input signal applied to a single input channel can be controlled by a voltage applied to the gate terminal, or, when signals are applied to both input channels, selection between them can be accomplished by switching the voltage applied to the gate terminal.

Referring to figure 4.7 the operation of the device is as follows; transistor T_7 acts as a constant current source and this constant current is divided between the two differential input channels consisting of transistor pairs T_1, T_2 and T_3, T_4 by a switching differential amplifier T_5, T_6. The switching differential amplifier is controlled by a signal applied to the gate terminal. If the gate terminal is left open or is biased sufficiently positive, T_5 is 'on', T_6 is 'off' and the constant current passes through the differential amplifier T_3, T_4. On the other hand if the gate terminal is connected to the negative supply T_5 is 'off', T_6 is 'on' and the constant current passes through the differential amplifier T_1, T_2. The collectors of the transistors in both channels are connected together to common load resistors ($1k\Omega$). The arrangement ensures that if zero signals are applied to the two input channels (and offsets balanced), the current flowing through each of the $1k\Omega$ load resistors is constant and does not depend upon which channel is activated. This means that there is very little d.c. level shift at the output when one channel is turned off and the other channel is turned on. The device has a low output impedance provided by the emitter followers which are used to supply the output signal to the differential output terminals.

4.4 Balanced Modulators

Cross coupled differential amplifiers are used in several linear i.c.'s to give devices capable of a four quadrant multiplier operation, a four quadrant capability allows the devices to be used as balanced modulators. A circuit model which we use to analyse the basic operation of such devices is shown in figure 4.8. The

Linear Integrated Circuit Applications

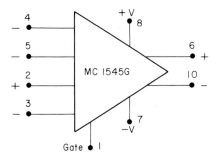

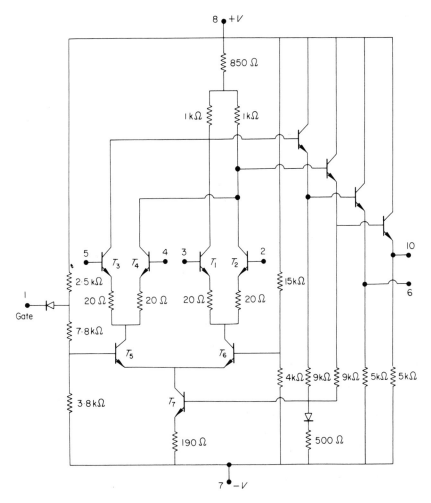

Fig. 4.7　Circuit diagram for MC 1545G, gate controlled two-channel input wide band amplifier

130   Linear Integrated Circuit Applications

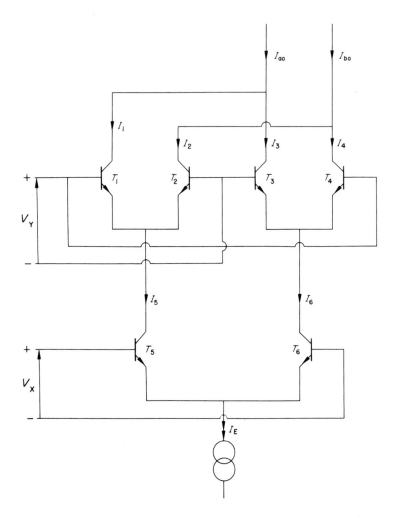

Fig. 4.8   Basic circuit model of variable transconductance balanced modulator

model consists of two cross coupled differential amplifiers composed of transistor pairs T_1, T_2 and T_3, T_4 with a third differential amplifier used to divide a constant current I_E between them. We assume that differential input signals are restricted to small values, ($V_{\text{diff}} < V_T$) which allows us to use approximate relationships previously introduced as equations 4.6. and 4.7.

Using these relationships we can write

$$\left. \begin{array}{ll} I_1 = I_5 \left(0.5 + \dfrac{1}{4V_T} V_Y\right), & I_2 = I_5 \left(0.5 - \dfrac{1}{4V_T} V_Y\right) \\[2mm] I_3 = I_6 \left(0.5 - \dfrac{1}{4V_T} V_Y\right), & I_4 = I_6 \left(0.5 + \dfrac{1}{4V_T} V_Y\right) \\[2mm] I_5 = I_E \left(0.5 + \dfrac{1}{4V_T} V_X\right), & I_6 = I_E \left(0.5 - \dfrac{1}{4V_T} V_X\right) \end{array} \right\} \quad (4.9)$$

Now $I_{ao} = I_1 + I_3$

and $I_{bo} = I_2 + I_4$

Substitution and simplification gives

$$I_{ao} = I_e \left(0.5 + \dfrac{1}{8 V_T^2} V_X V_Y\right)$$

and $I_{bo} = I_e \left(0.5 - \dfrac{1}{8 V_T^2} V_X V_Y\right)$

The differential output current

$$I_{ao} - I_{bo} = \dfrac{1}{4 V_T^2} V_X V_Y \quad (4.10)$$

The sign of the differential output current is dependent upon the sign of both V_X and V_Y, a four quadrant multiplier operation is performed by the circuit but linear operation is only possible for small values of the differential input signals. Note that the common mode level at the Y input channel must be greater than that at the X input.

4.4.1 The Balanced Modulator type SG 1402, (Silicon General)

The variable gain, wide band amplifier/multiplier type SG 1402 is an example of a device which employs the circuit principles outlined in figure 4.8, a circuit schematic for this linear i.c. is given in figure 4.9. In this schematic the transistor differential pairs Q_5 Q_7, Q_9 Q_{12} and Q_6 Q_{10} are used to perform the multiplier operation. Additional circuitry is incorporated in the device to provide appropriate biasing conditions and to convert the differential output current of the multiplier circuitry into a low impedance differential output voltage. The device is primarily intended for a.c. applications, including, variable gain amplification, modulation and demodulation; the bandwidth is said to extend to 50 MHz.

Schematic diagram

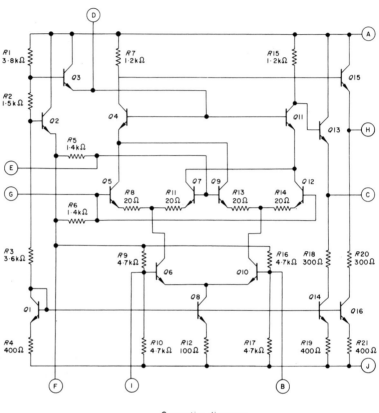

Connection diagrams

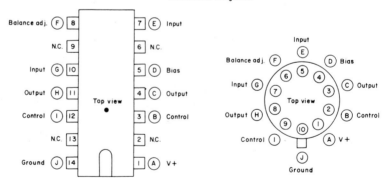

Fig. 4.9　　Balanced modulator, type SG 1402

4.4.2. The Balanced Modulator type MC 1596.

Another example of a linear i.c. which employs variable transconductance differential amplifier techniques is the balanced modulator type MC 1596,

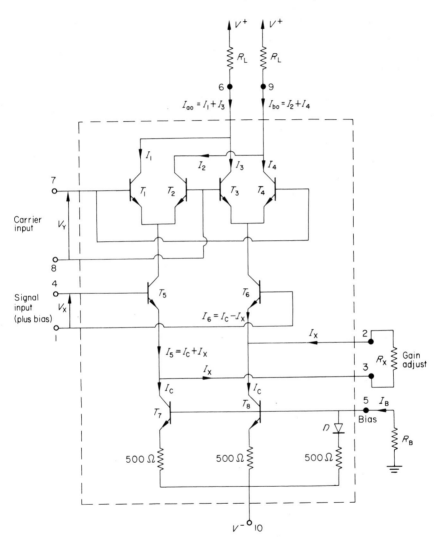

Components outside dotted line are connected externally

Fig. 4.10   Circuit schematic for balanced modulator, type MC 1596G

(Motorola, Signetics type N5596), a circuit schematic for the device is given in figure 4.10. As can be seen it requires external biasing components and its differential output current is converted into a differential output voltage by the connection of appropriate external load resistors. A choice of value for the external resistor R_X allows the designer to control the amplitude of the signal which can be handled linearly at the X input, R_X also determines the scaling factor or 'gain' of the device. A requirement for several external components may in one sense be regarded as a disadvantage but it does make the device versatile and it allows the designer to choose operating conditions appropriate to the complete system in which the modulating device may just be a single element.

The circuit operating principles of the 1596 are sufficiently different from those of the model in figure 4.8 to warrant a brief analysis. Instead of a single constant current source the device uses two current sources which are provided by transistors T_7 and T_8; temperature stabilisation of the currents is achieved by the method previously described in Section 4.2. using diode D. The two current sources are equal, they set a current I_c, where $I_c = I_b$ and I_b is the bias current which is supplied externally to pin 5 of the device. The value of the bias current I_b is fixed by choice of the resistor R_B and is determined by the relationship

$$I_B = \frac{|V_S| - V_D}{R_B + 500} \qquad (4.11)$$

V_D is the diode forward voltage, it has a value of approximately 0.75 V. at 25°C.

As in the previous analysis we assume that the differential input voltage V_Y is less than V_T, the expressions for I_1, I_2, I_3 and I_4 are thus the same as in equations 4.9 but the currents I_5 and I_6 are now

$$I_5 = I_C + I_X$$

and $I_6 = I_C - I_X$

As before we have, $I_{a0} = I_1 + I_3, I_{bo} = I_2 + I_4$ and a substitution and simplification gives

$$I_{ao} = I_c + \frac{1}{2 V_T} I_X V_Y$$

$$I_{bo} = I_c - \frac{1}{2 V_T} I_X V_Y \qquad (4.12)$$

The differential output current (for $V_Y < V_T$), is

$$I_{ao} - I_{bo} = \frac{1}{V_T} I_X V_Y \qquad (4.13)$$

The normal requirement is for a linear relationship between I_X and the differential input voltage V_X. If pins 2 and 3 are shorted together, making $R_X = 0$, the circuit for the 1596 becomes essentially the same as that of our model in figure 4.8. Linear operation then requires that $V_X < V_T$. Under these conditions

$$I_X = \frac{1}{4 V_T} 2 I_C V_X, \text{ from equation 4.8}$$

and equation 4.13

becomes $\quad I_{ao} - I_{bo} = \dfrac{I_C}{2 V_T^2} V_X V_Y \qquad (4.14)$

$(R_X = 0, V_Y \text{ and } V_X < V_T)]$

With $\quad I_E = 2I_C$ equation 4.14. is identical to equation 4.10
A resistor R_X connected between pins 2 and 3 effectively introduces negative feedback at the X input. The differential input voltage, V_X, in addition to supplying a differential base emitter voltage between transistors T_5 and T_6 must now supply a voltage drop $I_X R_X$. Thus

$$V_X = I_X R_X + (V_{BE5} - V_{BE6}) \qquad (4.15)$$

The differential base emitter voltage varies non-linearly with I_x but, provided $I_x < I_{01}(V_{BE5} - V_{BE6})$ never becomes appreciably greater than V_T (see figure 4.2). If the value of R_x is chosen so that $R_x \gg V_T/I_c$ then $I_x R_x \gg (V_{BE5} - V_{BE6})$. Under these conditions we can write equation 4.15 as

$V_x \triangleq I_x R_x$, and equation 4.13 becomes

$$I_{ao} - I_{bo} = \frac{1}{V_T R_x} V_Y V_x \qquad (4.16)$$

$(R_x \gg \dfrac{V_T}{I_c}, V_{x\,max} < I_c R_x, V_Y < V_T)$

The inclusion of the resistor R_x removes the small signal limitation at the X input. With R_x suitably chosen the maximum signal which can be handled linearly at the X input is determined by the relative voltage bias levels which are externally applied to the two input channels and to pin 10. The signal amplitude which can be handled linearly at the Y input is limited to values less than V_T. If

$V_Y \gg V_T$ the Y input effectively acts as a switching signal which switches the current I_x between the differential output terminals. The differential output current, $I_{ao} - I_{bo}$, is switched between the values $\pm 2I_x$, as V_Y takes on its positive and negative values respectively.

In all modes of operation expressions for the differential output voltage are simply obtained by multiplying the appropriate expression for the differential output current by the magnitude of the external load resistor, R_L.

4.5 Variable Transconductance Linear Multipliers

Both the modulating devices discussed in the previous section have differing d.c. levels at their two input channels and this, together with small signal limitations, makes the devices unsuitable for multiplier operations involving d.c. input signals referred to a common earth. In figure 4.11 we show a simplified circuit schematic for a variable transconductance multiplier cell in which the difficulties are largely overcome[2,3]. It consists essentially of the circuitry of figure 4.10, (the MC 1596 device), plus additional components which are used to process the Y input signal in such a way as to remove the small signal limitations on this channel and to shift its d.c. level. The similarities between the two circuits allows us to analyse the circuit of figure 4.11 as a development of the analysis previously carried out for figure 4.10 (section 4.4.2).

Using equation 4.15 we write

$$I_x \simeq \frac{V_x}{R_x} \qquad (4.17)$$

conditions;

$$I_x < I_{cx}, \quad R_x \gg \frac{V_T}{I_{cx}}$$

$$I_Y \simeq \frac{V_Y}{R_Y} \qquad (4.18)$$

conditions;

$$I_Y < I_{cY}, \quad R_Y \gg \frac{V_T}{I_{cY}}$$

Using equation 4.11 we write

$$I_{cx} = I_{Bx} = \frac{|V_s^-| - V_D}{R_{Bx} + R} \qquad (4.19)$$

$$I_{cY} = I_{BY} = \frac{|V_s^-| - V_d}{R_{BY} + R} \qquad (4.20)$$

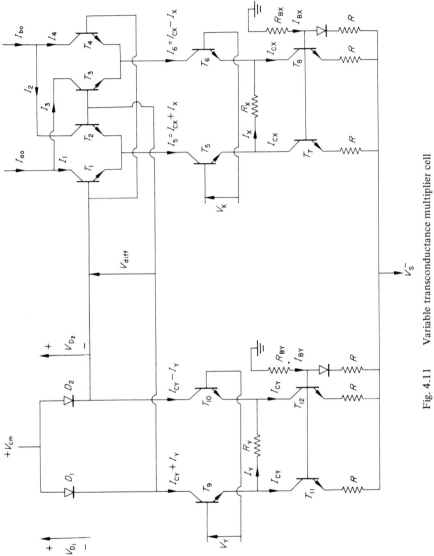

Fig. 4.11   Variable transconductance multiplier cell

The diode voltage drop V_D is approximately 0.75 V at 25°C.

We now make use of equations 4.4 and 4.5 in order to write expressions for the currents I_1, I_2, I_3 and I_4. As in previous analyses we assume matched transistors and that transistor base currents may be neglected compared to their collector currents.

Thus

$$I_1 = \frac{I_{cx} + I_x}{1 + \exp - \frac{V_{diff}}{V_T}} \qquad I_2 = \frac{I_{cx} + I_x}{1 + \exp + \frac{V_{diff}}{V_T}}$$

$$I_3 = \frac{I_{cx} - I_x}{1 + \exp + \frac{V_{diff}}{V_T}} \qquad I_4 = \frac{I_{cx} - I_x}{1 + \exp - \frac{V_{diff}}{V_T}} \qquad (4.21)$$

Now

$$I_{ao} = I_1 + I_3$$

and

$$I_{bo} = I_2 + I_4$$

substituting values from equations 4.21 and simplifying we obtain

$$I_{ao} - I_{bo} = 2 I_x \left[\frac{1}{1 + \exp - \frac{V_{diff}}{V_T}} - \frac{1}{1 + \exp + \frac{V_{diff}}{V_T}} \right] \quad (4.22)$$

We now note that

$$V_{diff} = V_{cm} - V_{D2} - (V_{cm} - V_{D1})$$
$$= V_{D1} - V_{D2}$$

Diode D_1 carries the current $I_{cY} + I_Y$ and diode D_2 carries the current $I_{cY} - I_Y$. Provided that the current through either diode does not fall below a value which makes the diode forward voltage less than approximately 100 mV we can make use of the approximated diode equation, $I = I_0 \exp V_0/V_T$ (similar to equation 4.2) and write

$$I_{cY} + I_Y = I_0 \exp \frac{V_{D1}}{V_T}, \quad I_{cY} - I_Y = I_0 \exp \frac{V_{D2}}{V_T}$$

Thus

$$\frac{I_{cY} + I_Y}{I_{cY} - I_Y} = \exp \frac{V_{D1} - V_{D2}}{V_T} = \exp \frac{V_{diff}}{V_T}$$

Substitution for the exponential term in equation 4.22 gives

$$I_{ao} - I_{bo} = \frac{2}{I_{cY}} I_x I_Y \qquad (4.23)$$

and substituting the values of I_x and I_Y from equations 4.17 and 4.18 we further obtain

$$I_{ao} - I_{bo} = \frac{2}{I_{cY} R_x R_Y} V_x V_Y \qquad (4.24)$$

If equal load resistors R_L are used to convert the differential output current into a differential output voltage V_0

$$V_0 = \frac{2 R_L}{I_{cY} R_x R_Y} V_x V_Y \qquad (4.25)$$

The quantity

$$K = \frac{2 R_L}{I_{cY} R_x R_Y}$$

is the so-called scaling factor of the multiplier, the practical considerations involved in the choice of scale setting components will be considered later.

The circuit model of figure 4.11 shows essentially the circuitry which is included in the MC 1495/1595 multiplying device. The MC 1495/1595 was the first commercially available integrated circuit capable of performing a four quadrant linear multiplication. It is a basic multiplying cell and it requires many external components to make it work in a practical application. It provides an output signal in the form of a differential current which must be converted into a differential voltage by two equal resistors connected externally between the device and the positive supply rail. A normal requirement for a multiplier is that it give a single-ended output voltage which is referenced to earth so that additional circuitry is required with the 1595 in order to shift the level of the output signal and convert it into a single-ended form.

There are now several other variable transconductance linear four quadrant multipliers available in monolithic form, they use basically the same circuit operating principles as the 1595 multiplier cell but have added circuitry on the device chip which allows them to be used with fewer external components or gives them a greater application versatility. The MC 1494/1594 is an improved version of the 1595, it includes a current and voltage regulator and a differential current converter. The Analog Devices monolithic multiplier type AD 530, AD 531, AD 532, Intronics Type M530, Barr Brown Type BB 4203, all include

140 Linear Integrated Circuit Applications

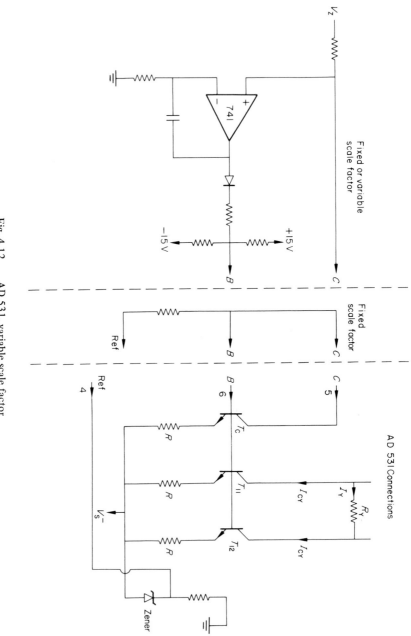

Fig. 4.12 AD 531, variable scale factor

an output operational amplifier on the device chip; these devices require fewer external components than the 1594 and are simpler to use. Analog Devices type AD 531 multiplier includes an output operational amplifier and in addition it allows for a scale factor that can be dynamically varied in such a way as to provide an overall transfer function of the type $V_x V_Y/V_z^4$.

Referring to the basic multiplier circuit model of figure 4.11 the AD 531 achieves a dynamically adjustable scale factor by an external control of the current I_{cY}, (see equation 4.25). The circuit modification which is used to perform the control function is shown in figure 4.12. In applications requiring a fixed scaling factor, pins 5 and 6 are connected together and joined by an external current setting resistor to the reference voltage at pin 4, alternatively the scaling current, $(I_{cY} = I_{bY})$, is derived by connecting pins 5 and 6 to earth through an appropriate resistor when the circuit is then essentially the same as figure 4.11. In applications requiring a variable scale factor the control transistor, T_c, derives its collector current, I_{BY}, from a controlled current source in the form of the externally connected voltage to current op. amp. circuit shown in figure 4.12. The control transistor is connected within the feedback loop of the op. amp.; feedback is returned to the non-inverting input terminal of the amplifier because of the phase inversion between base and collecter of the transistor. A resistive network is used to appropriately shift the output voltage level of the operational amplifier.

It seems likely that many more monolithic multiplying devices will shortly be commercially available. In making his choice of a particular device the user should carefully assess his performance requirements and when making cost comparisons should not forget the external components that the device will need nor the time it will take to adjust it so that it will achieve its rated performance. Practical considerations involved in using multipliers and some of their many applications will be dealt with in chapter 6.

References

1. A.B. Grebene, *Analog Integrated Circuit Design*, Van Nostrand Reinhold (1972)

2. B. Gilbert, *A Precise Four-Quadrant Multiplier with Subnanosecond Response*, I.EEE, *J. Solid State Ccts. SC* – 3 (4) (1968) 365-73.

3. E. Renschler, *Theory and Application of a Linear Four-Quadrant Multiplier,* EEE Magazine, 15 (**5**) (May 1969).

4. Analog Devices, *AD531 Data Sheet.*

Exercises 4

4.1 The 'long tailed pair' amplifier is connected as shown in figure 4.3 with the following component values: $R_B = 22$ kΩ $R_{L1} = R_{L2} = 10$ kΩ, $V_s = \pm 6$ volts. Assuming perfect transistor matching, what is the common mode output voltage level at the collectors of the differential pair when both input terminals of the amplifier are earthed?
What are the values of the collector voltages if one input terminal is earthed and a voltage of 13 mV with respect to earth is applied to the other input terminal?

4.2 The balanced modulator device type 1596, shown in figure 4.10 has the following values in the circuit: $R_x = 5$ kΩ, $R_b = 15$ kΩ $R_{L1} = R_{L2} = 10$ kΩ, $V_s = \pm 10$ volts. Write down the relationship between the differential output voltage and the differential input signals V_X and V_Y. Consider the limitations on V_X and V_Y for linear operation.

5. Variable Gain Devices — Practical Considerations

Chapter 4 was devoted to a consideration of the basic circuit principles underlying the action of integrated circuit variable gain devices, this chapter introduces some of the practical capabilities of these devices. The treatment to be given is by no means exhaustive and if more detailed information is required about the characteristics and capabilities of a particular device reference should be made to the appropriate manufacturers data sheets.

Figure 5.1 illustrates a practical circuit which is an example of an application of devices of the type analysed in section 4.2. The cascode arrangement (common emitter, common base), possible with this kind of device, features a very small interaction between output and input making the devices very useful for tuned RF and wide band amplifier applications[1,2].

5.1 Controlled Operational Amplifiers (Programmable Operational Amplifiers)

In most applications the practical considerations involved in using a programmable op. amp. are not significantly different from those involved in using any other operational amplifier type. For example, the Fairchild programmable op. amp., type μA776 can, by choice of bias setting resistor, be given characteristics similar to those of the 741 device used in the applications given in earlier chapters. The pin configuration for the μA776 is similar to that of the 741 and is shown in figure 5.2. Note that the 'set' current flows out of pin 8 and the bias setting resistor is connected to earth or the negative supply rail.

By a suitable choice of bias setting resistor the μA776 can be used with power supply voltages as low as ± 1.2 V with the quiescent power consumption down to a few microwatts or less. A practical circuit which illustrates the low power consumption capabilities of the μA776 is shown in figure 5.3, it is an inverting feedback amplifier and it is said to operate with a power consumption of only 600 nW. In addition to controlling the power consumption of the device the set current also determines the values of other important amplifier parameters and thus allows the characteristics of the

144   *Linear Integrated Circuit Applications*

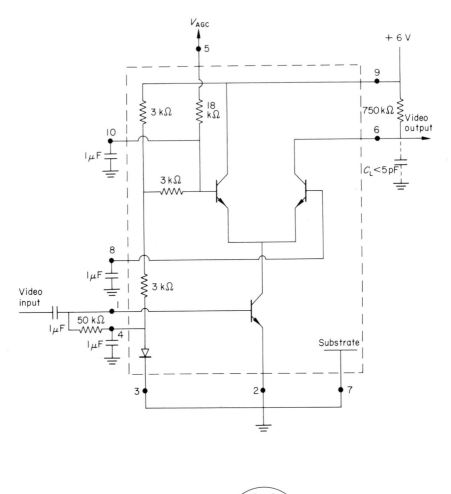

Pin 7 connected to case

Fig. 5.1   Circuit schematic for RF/IF amplifier, type MC 1550G, connected as video amplifier. Circuit has bandwidth 20MHz. With $V_{AGC} = 0$, gain = 25dB. With $V_{AGC} = +4V$, gain = 12dB; bandwidth remains constant over AGC range

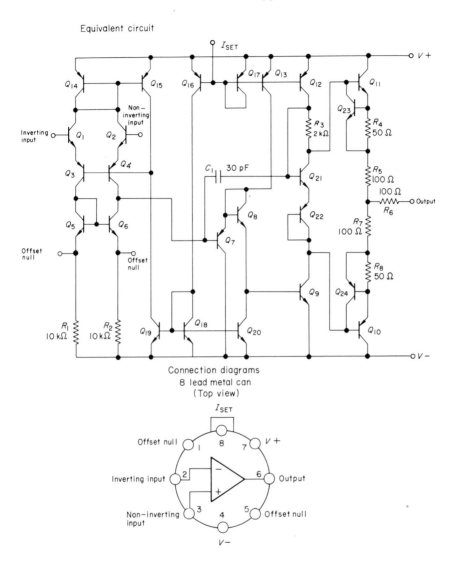

Fig. 5.2   Fairchild programmable operational amplifier, type μA776

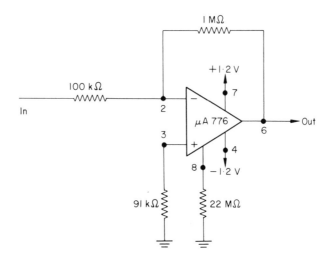

Fig. 5.3 Inverting amplifier with very low power consumption

amplifier to be adjusted to values which best suit the requirements of a particular application.

Control capabilities can be exploited to allow applications that are not directly possible with more conventional op. amps., the circuit in figure 5.4 is given as an example. The function of the circuit, which is that of a three channel multiplexer using the triple operational transconductance amplifier array type CA 3060, is to allow transmission of any one of the three input signals, e_1, e_2 and e_3. Selection is accomplished by the application of a bias current to a particular amplifier, only that amplifier which receives a bias . current is operative and transmits the input signal applied to it. The output signal is provided by the 3N138 M.O.S.F.E.T. which is used as a buffer and power amplifier and is connected within the feedback loop. The M.O.S.F.E.T. produces a phase inversion and the negative feedback condition is thus satisfied by a feedback path from the output to the non-phase inverting terminal of each op. amp. The cascade arrangement of each amplifier with the M.O.S.F.E.T. provides a large loop gain, (of an order 100 dB), and gives the circuit the desirable characteristics normally associated with the follower configuration; high input impedance, low output impedance, etc. . The series connected 390 Ω resistor and 1000 pF capacitor are used to provide frequency compensation and ensure closed loop stability.

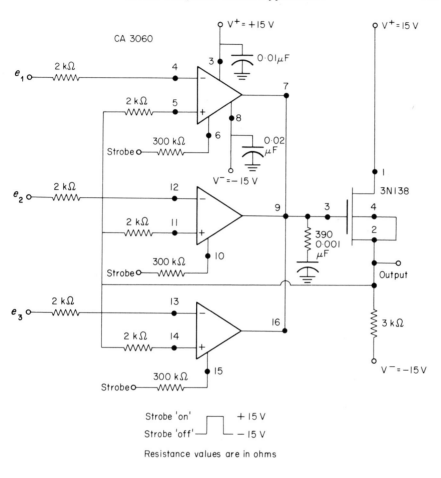

Fig. 5.4   Three-channel multiplexer using triple amplifier type CA 3060 (RCA)

The manufacturer's application notes suggest a wide variety of different applications for the CA3060 device including a circuit for a four quadrant multiplier[3]. It would seem however that multiplier/modulator applications are more conveniently implemented using i.c. devices which are specifically designed for this purpose. Indeed, when there are several devices available which are all apparently capable of performing a desired function it is often quite difficult to decide which is the best one to use. The choice can only be made with a full knowledge of the complete system requirements. Clearly if a system requires the use of several op. amps. and there is the further requirement that power consumption be small then the use of a device like the CA3060 or the L144, (Silconix), should be seriously considered.

5.2 The Two Channel Gate Controlled Wide Band Amplifier, Type 1545

The circuit operating principles of the 1545 device were discussed in section 4.3.2. It is a somewhat unusual device in that it is a wide band amplifier with two differential input channels, a single-ended control channel and a differential output channel. Parameters are required to describe the electrical relationships which exist between the signals at the various channels; experimental procedures for the measurement of some of the basic device characteristics are now given. A practical familiarity gained by performing these experiments should make the reader more readily appreciate the significance of the full list of performance parameters which are specified on the device data sheet.

5.2.1 *Parameter Measurements for the 1545 Device*

Input Offsets and Output Level Shifts

A test circuit is shown in figure 5.5. Voltage measurements in the tests should be made with a high impedance d.c. millivoltmeter. Referring to figure 5.5, the channel with input terminals 4 and 5 is activated first, this is done by setting the switches to position (a). A voltage derived from the potentiometer is applied to pin 4 and adjusted to balance out the input offset voltage and make the differential output voltage, V_{10-6} zero. The input offset voltage associated with the input channel 4, 5, is then read directly at pin 4. The quiescent output voltage level at pin 6 should also be recorded. Some readings obtained with this test circuit are given; $V_4 = V_{io_{4,5}} = 2.4$ mV, $V_6 = V_{10} = 0.51$ V
The procedure is repeated for the other channel with the switches set to position (b). Readings; $V_2 = V_{io_{2,3}} = 0.2$ mV, $V_6 = 0.49$ V. Note that for the particular device tested switching between channels causes a single-ended output level change of 20 mV.

Input bias currents may be determined by connecting 1 kΩ resistors, (ideally matched and selected to say 1 per cent), to the appropriate input terminals and measuring the voltage produced by the bias current flowing through these resistors. Some readings: gate set at + 5 V, channel 4, 5, activated;

$$V_4 = 14.9 \text{ mV}, \quad V_5 = 14.6 \text{ mV}$$

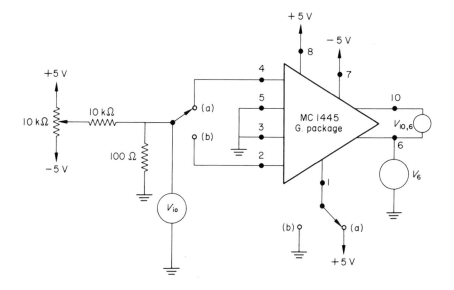

Fig. 5.5   Test circuit for measurement of input offsets and output level shifts

giving

$$I_4 = 14.9\ \mu A, \quad I_5 = 14.6\ \mu A$$

and

$$I_{io_{4,5}} = 0.3\ \mu A$$

Gate set at 0 V, channel 2.3 activated

$$V_2 = 15.1\ mV, \quad V_3 = 15\ mV$$

giving

$$I_2 = 15.1\ \mu A, \quad I_3 = 15\ \mu A$$

and

$$I_{io_{2,3}} = 0.1\ \mu A.$$

Note that all device parameter values are subject to a considerable production spread. The numerical values given refer to a specific device tested by the author, they lie within the manufacturers quoted spread but they

150   Linear Integrated Circuit Applications

should not necessarily be regarded as typical values. Indeed the value of 'typical' parameter magnitudes, that are often quoted on manufacturers data sheets, seems somewhat dubious from the designers point of view. The designer must really know the possible maximum and minimum values of relevant device parameters if he is sensibly to design for a specific performance requirement.

Gate Control Characteristic

A test circuit for the measurement of the gate control characteristic of the 1545 is shown in figure 5.6 A sinusoidal signal of fixed amplitude and frequency (say 20 mV, 10 kHz), is applied to the input terminal pin 2, and the single-ended output voltage at pin 6 is measured for different values of the gate d.c. voltage. Some results obtained with this test are illustrated graphically in figure 5.7; the single-ended voltage gain is shown plotted against the gate control voltage. Note that the gain varies almost linearly with the gate control voltage over part of this characteristic.

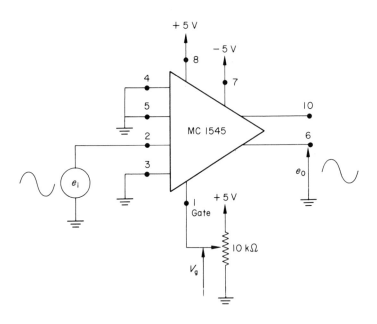

Fig. 5.6   Measurement of gate control characteristic

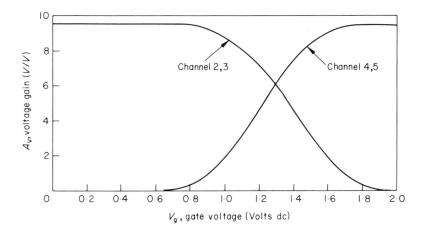

Fig. 5.7 Single-ended gain as a function of gate voltage (MC 1545)

5.2.2 Applications of the 1545 Device

The 1545 is a versatile device which can be applied in a variety of different ways, but, because a device can perform a specific function it does not mean that it is necessarily the best or most convenient one to use for that function. As mentioned previously the decision as to which device to use in a specific situation can only be made with a complete knowledge of the system requirements. The gate control channel of the 1545 allows the device to be used as an amplitude modulator and by cross-coupling the differential input channels it can be used as a balanced modulator. Modulator applications are probably more conveniently performed by modulator devices like the SG 1402 or MC 1596 and are therefore not discussed in detail here. Examples of applications that can be performed particularly conveniently with the 1545 (perhaps not so conveniently with an alternative device), are given. The reader is referred elsewhere[4,5] for a more comprehensive selection of the possible applications of the 1545.

Wide Band Amplifier with AGC

One of the basic applications of the device is that of a wide band amplifier. In such an application the input signal is applied to only one of the two input channels, the input terminals of the channel not in use are connected to earth and a d.c. voltage is applied to the gate terminal in order to control the gain. The input signal may be single-ended or differential and a single-ended output signal may be taken from pin 6 or a differential output signal from between pins 10 and 6; the output impedance is low and the output signal is referenced close to earth. Maximum gain is of the order 20 dB and the bandwidth extends to typically 75 MHz.

Analogue Switch

The MC1545 may be used as an analogue switch controlled by a digital logic signal applied to the gate terminal; this application is illustrated in figure 5.8. In this an input signal is applied to pin 4, this signal is amplified and appears at the output when the control signal applied to the gate is high, corresponding to a logic state 1. The signal is not passed through the amplifier when the gate control signal is low, (logic state 0). If it is required that the opposite logic state pass or block the signal, the signal should be applied to pin 2 or 3 and pins 4 and 5 should be earthed. The circuit supplying the gate control signal must sink a maximum of 2.5 mA in the low state; in the high state it need only source the leakage current of a reverse biased diode. These requirements are compatible with most digital logic circuits.

Multiplexing

In Figure 5.8 only one of the input channels of the device is in use; if rather than earthing the second channel a second signal is applied to it the

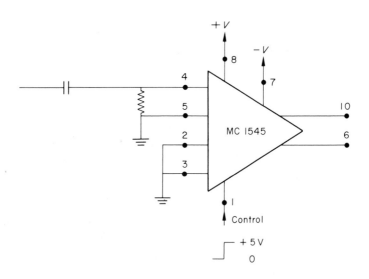

Fig. 5.8　　MC 1545 used as an analogue switch

gate control signal acts as a channel select signal. The principle can be extended by paralleling and cascading several 1545 units to allow selection of one out of 'n' inputs. An example in the form of a one-out-of-four selector is illustrated in figure 5.9. The control signals which must be applied in order to select a particular input are shown in the accompanying input/output truth table.

Gated Oscillator

The wide bandiwdth of the MC1545 allows it to be used as an oscillator at much higher frequencies than would be possible using an op. amp. as the active device in the circuit. In addition the gate control signal provides a

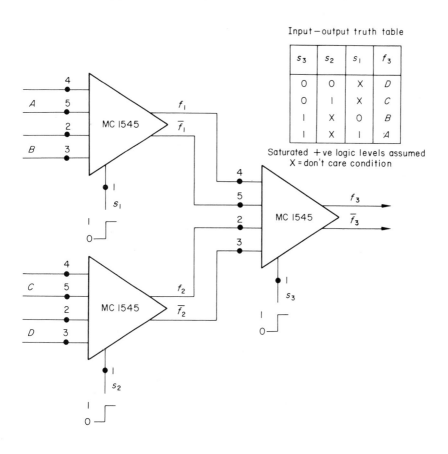

Fig. 5.9   One-out-of-four selector

154   Linear Integrated Circuit Applications

convenient method of starting and stopping oscillations. A circuit for a gated oscillator is shown in figure 5.10; it uses only one of the input channels of the MC1545 and is essentially very similar to the op. amp. Wien Bridge oscillator. The frequency of oscillations is determined by the values of the resistors, R and the capacitors C and may be calculated from the equation,
$f = 1/2 \pi C R$; the useful frequency range is said to extend from below 1kHz to 10 MHz. The oscillations are gated-on by a positive voltage applied to pin 1 and take a finite time to reach their final amplitude, this time varies from 1 to several cycles of the oscillation frequency and is dependent upon $C R$ values. Oscillations stop in a time equal to the channel select time (20 ns), when the gate voltage is switched to zero. The amplitude of the oscillations is stabilised by the diode resistor network. The network, which is connected in the

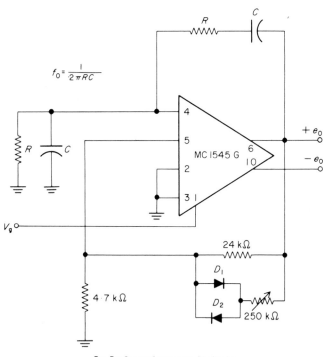

Fig. 5.10   Gated oscillator using MC 1545

negative feedback path, makes the magnitude of the loop gain unity when the oscillation amplitude reaches its required value.

5.3 Modulator Applications

Signal modulation and demodulation are processes which essentially involve a multiplication operation and any device which produces an output signal which is proportional to the product of two input signals can, in principle, be used for modulator applications. Most of the devices which were considered in chapter 4 fall in this category but some are more suitable than others for modulator applications. Before considering practical modulator circuits brief theoretical analyses of the various modulation processes are given.

5.3.1 Modulation Processes

Balanced Modulation
Consider two constant amplitude pure sine wave signals

$$e_{1(t)} = E_1 \cos(\omega_1 t + \phi) \text{ and } e_{2(t)} = E_2 \cos \omega_2 t.$$

The product

$$e_{1(t)} \, e_{2(t)} = \frac{E_1 E_2}{2} \left[\cos[(\omega_1 + \omega_2) t + \phi] + \cos[(\omega_1 - \omega_2) t + \phi] \right] \quad (5.1)$$

The result of linearly multiplying together two sinusoidal signals is to give a product which consists of two components which are sinusoidal and have frequencies which are the sum and difference frequencies of the two signals used to form the product, the product contains no other frequency components. The process is called balanced modulation, in communications systems it is referred to as 'suppressed carrier double sideband modulation' (S.C.D.S.B.). In such systems $\omega_1 = \omega_{c1}$ the carrier frequency and $\omega_2 = \omega_m$ the modulation frequency. If only one of the two frequency components in the product is required the other may be filtered out, for example, if ω_1 and ω_2 are close together a simple low pass filter may be used to reject the frequency sum leaving only the difference frequency term.

Amplitude Modulation
Consider the effect of adding a d.c. term to one of the signals used to form the product, the two signals being

$$e_{1(t)} = E_1 \cos \omega_c t \text{ and } e_{2(t)} = E_2 (1 + m \cos \omega_m t)$$

The product is

$$e_{1(t)} \, e_{2(t)} = E_1 E_2 \cos \omega_c t + \frac{m}{2} E_1 E_2 [\cos(\omega_c + \omega_m) t + \cos(\omega_c - \omega_m) t] \quad (5.2)$$

The product contains three frequency components, one at the carrier frequency the other two at the carrier and modulation sum and difference frequencies. The process is called amplitude modulation.

Frequency Doubling

A multiplication process may be used to produce frequency doubling, if the two sinusoidal signals which are multiplied together are identical ($e_{(t)} = E \cos \omega t$), the product

$$e_{(t)}^2 = E^2 \cos^2 \omega t$$

$$= \frac{E^2}{2} (1 + \cos 2 \omega t) \tag{5.3}$$

The multiplication process gives rise to a d.c. term equal to one half the peak voltage squared and an alternating signal at twice the frequency of the signal $e_{(t)}$.

Phase Sensitive Detection, PSD

The multiplication of two sinusoidals signals, $e_{1(t)} = E_1 \cos (\omega t + \phi)$ and $e_{2(t)} = E_2 \cos \omega t$, of the same frequency but with a phase difference ϕ gives the product

$$e_{1(t)} e_{2(t)} = \frac{E_1 E_2}{2} (\cos 2 \omega t + \cos \phi) \tag{5.4}$$

The product consists of an alternating component, which can be easily filtered out by a low pass filter, and a d.c. term which is proportional to the cosine of the phase difference between the two signals used to form the product.

Demodulation

If an amplitude modulated carrier signal is multiplied by a signal at the carrier frequency the carrier forms sum and difference frequency components with the components of the amplitude modulated signal. This process gives rise to a component at the modulation frequency in the product and if $\omega_c \gg \omega_m$, a simple low pass filter may be used to reject the other components; the modulating signal is in this way recovered from the modulated carrier.

Effect of Non-Linearities

A repetetive non-sinusoidal signal can be considered as a superposition of a sinusoidal signal and a series of harmonics. The result of multiplying together two non-sinusoidal signals is to give a product which contains sinusoidal components at the sum and difference frequencies of all the frequency components present in the two signals used to form the product. A practical multiplier which is not linear, effectively distorts the signals applied to its input channels. A multiplication of two sinusoidal signals in this case give rise to an output signal which contains components at the sum and difference frequencies of all the harmonics of the two signals which are introduced as a result of the distortion.

5.3.2 Modulator Circuits

Two of the devices discussed in chapter 4 are particularly suitable for modulator applications, the SG 1402 and the MC 1596. Four quadrant linear multiplier devices also readily lend themselves to modulator applications. Bandwidth, signal handling capability, linearity, nature of the output signal, number of external components required, compatability with the other elements in the system and, of course, cost are all factors which require consideration when choosing a modulating device for a specific application.

Of the devices mentioned, the MC 1596 is capable of the highest frequency operation, it may be used in modulator demodulator applications at frequencies up to 100 MHz. Operating conditions in the MC 1596 are determined by the values of external components and this allows the designer to control some of the device characteristics. Biasing conditions are fixed internally in the SG 1402 device; this means that fewer external components are required but the designer has less control over the device characteristics. Four quadrant linear multipliers allow a linear operation for a wider range of input signal amplitudes than that possible with the two modulator devices but bandwidth considerations generally restrict their modulator/demodulator usage to the lower frequencies.

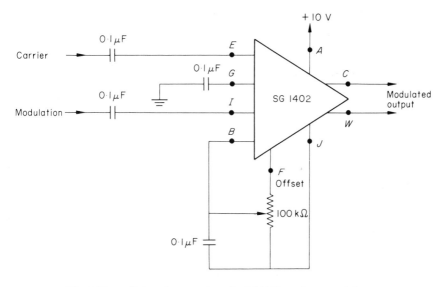

Fig. 5.11   External connections for SG1402 used as a modulator

A detailed internal circuit schematic for the SG 1402 device was given in figure 4.9; figure 5.11 shows the external connections that are required in order to use the device as a modulator. The 100kΩ potentiometer is used to apply an adjustable d.c. offset voltage to the differential input channel with input terminals I and B. Adjustment of the potentiometer allows the circuit to act either as a balanced modulator or as an amplitude modulator. If linear operation is required both input signal amplitudes must be restricted to small values. The amplitude of the signal applied to the input channel I, B, must be less than V_T, (see section 4.4), the signal amplitude at input terminals E, G, can be slightly larger because of the 20Ω emitter degeneration resistors associated with this input channel.

Using the Balanced Modulator Type MC 1596. Practical applications of the MC 1596 modulator device require attention to the external components which must be connected to the device. The external components, together with the externally applied voltages, are required to set the values of four bias voltage levels and a bias current. Referring to figure 5.12, which shows a typical biasing arrangement, the voltage levels which must be set are, in order, starting with the most positive

(1) The differential output terminals, pins 6 and 9, normally connected to the positive supply voltage by means of equal load resistors.
(2) The Y differential input channel, pins 7 and 8
(3) The X differential input channel, pins 1 and 4
(4) The most negative point in the circuit, pin 10; this terminal is connected to the negative supply voltage when using twin power supplies or to earth when using a single power supply.

Levels (2) and (3) may be set by resistive dividers or, when using twin power supplies, earth potential may be used as one level. The voltage levels chosen must ensure that the transistors do not saturate (allow at least 2 volts collector base voltage on all transistors), and that maximum power limitations are not exceeded (see device data sheet).

Operating currents in the MC 1596 are determined by the current sources I_C, (transistors T_7 and T_8). Note that $I_C = I_B$, where I_B is the externally supplied bias current controlled by the value of resistor R_B and determined by equation 4.11.

A final component which requires selection is the scale setting resistor R_X, the value to be used is dependent upon the operating current I_C and the expected maximum amplitude of the X input signal (see equation 4.16).

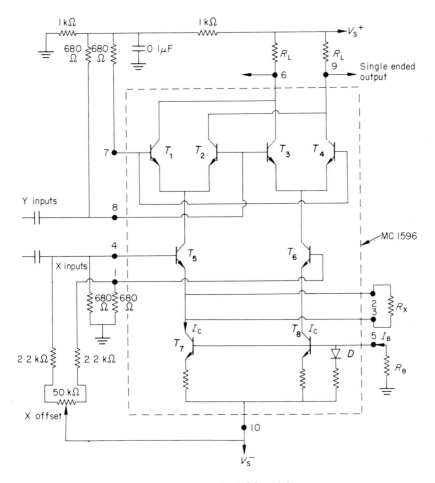

Fig. 5.12   Biasing of the 1596 modulator

The reader may find it instructive to make d.c. measurements of the terminal voltages in order to gain familiarity with the device operation. A variety of measurements can usefully be made, we suggest a simple experimental procedure as a basic guide; it is in no way a comprehensive procedure and the reader is encouraged to extend the scope of his measurements if possible.

Experimental Procedure (referring to figure 5.12):

With zero a.c. signal applied to the Y channel adjust the X offset potentiometer to make V_6 equal to V_9 and then measure V_6 and V_9 for a range of values of R_B. Some experimental results:

with $\quad R_L = 3.9\text{k}\Omega$, $R_X = 1\text{k}\Omega$, $V_{3+} = 12\text{V}$, $V_S^- = -8\text{V}$.

R_B	$V_6 = V_9$	$I_C = \dfrac{V_S^+ - V_6}{R_L}$	I_C (calculated from equation 4.11)
15kΩ	10.2 V	0.46 mA	0.47 mA
10kΩ	9.4 V	0.67 mA	0.69 mA
6.8kΩ	8.3 V	0.95 mA	0.95 mA

The dynamic behaviour of the circuit can be investigated in the following way. Apply an alternating signal of amplitude approximately 20mV to the Y channel at pin 8 and monitor input and the single-ended output signal at pin 6 or 9 whilst at the same time varying the d.c. offset voltage applied to the X channel. Note that as the amplitude of the output signal goes through a minimum its phase reverses. Set the offset potentiometer for minimum a.c. output, remove the alternating signal applied to the Y channel and apply a small d.c. offset to this channel. (Note, the bias current drawn by pins 7 and 8 can be made to apply on offset voltage to the Y channel if the resistors connected to pins 7 and 8 are made unequal). Apply an alternating signal to the X channel at pin 1 and observe the maximum amplitude of the X signal that can be applied before the output waveform shows evidence of distortion. Repeat the procedure using different values for the resistor R_B. Some experimental results which were obtained using a value of R_X equal 1kΩ are given:

R_B	$V_{X\,max}$	$I_{X\,max} = \dfrac{V_{X\,max}}{R_X}$	I_C
15kΩ	0.32 V	0.32 mA	0.46 mA
10kΩ	0.5 V	0.5 mA	0.67 mA
6.8kΩ	0.85 V	0.85 mA	0.95 mA

The results confirm that linear processing of the X input signal requires that $I_{X\,max}$ be less than the current I_C.

The circuit of figure 5.12 can be used as a balanced modulator; the carrier signal is applied to the Y channel at pin 8 and the X offset potentiometer is adjusted for a minimum (ideally zero), signal output; the modulating signal is now applied to the X channel at pin 4. The waveforms in figure 5.13 show a

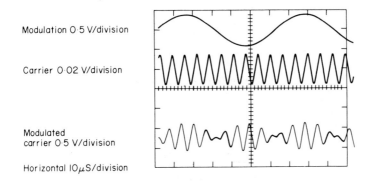

Fig. 5.13   Balanced modulation; linear operation

modulating signal, a carrier signal and the single-ended output signal produced at pin 6; the carrier signal amplitude is small and the circuit operates linearly. The waveforms in figure 5.14 show the effect of increasing the carrier signal amplitude to the point where the carrier is processed non-linearly. The frequency spectra of the output signals for linear and non-linear processing are shown in figure 5.15. Non-linear operation produces extra frequency components in the output signal as discussed in section 5.3.1. In applications in which the modulator is followed by a filter the spurious components are removed by the filter and operation with a high level carrier maximises gain and makes the output signal insensitive to small variations in carrier amplitude.

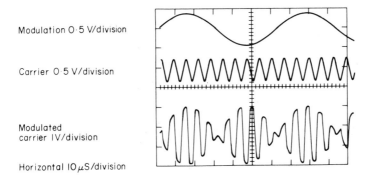

Fig. 5.14   Balanced modulation; non-linear operation

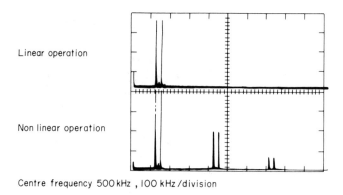

Fig. 5.15   Balanced modulation; frequency spectra

In order to use the circuit of figure 5.12 for amplitude modulation all that is required is to adjust the X offset potentiometer in order to apply a d.c. offset to the X channel. Waveforms illustrating the action of the circuit as an amplitude modulator are shown in figure 5.16, the potentiometer was adjusted to make $m = 1$. The frequency spectrum of the amplitude modulated output signal is shown in figure 5.17 and frequency spectra for high and low level carrier operation are compared in figure 5.18.

Readers requiring more detailed information about the MC 1596 are referred to manufacturers application notes [6].

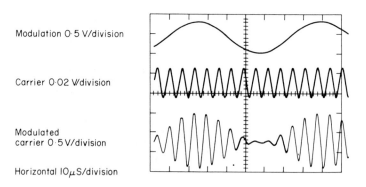

Fig. 5.16   Amplitude modulation, m = 1; linear operation

Linear Integrated Circuit Applications

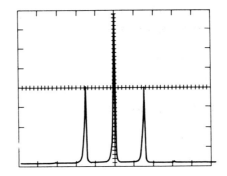

Centre frequency 150 kHz
10 kHz/division

Fig. 5.17   Amplitude modulation; frequency spectrum, linear operation; m = 1

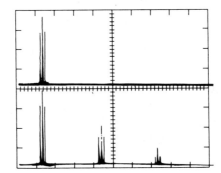

Centre frequency 500 kHz
100 kHz/division

Fig. 5.18   Amplitude modulation; frequency spectra for linear and non-linear operation compared

References

1. *An Integrated Circuit R.F./I.F. Amplifier*, Motorola, Appl. Note. *AN-247*

2. *An Integrated Circuit Wideband Video Amplifier with AGC*, Motorola Appl. Note *AN-299*

3. *Operational Transconductance Amplifier Arrays*, R.C.A., Linear Integrated Circuits File *No 537*

4. *Using the MC 1545 – A Monolithic Gated Video Amplifier*, Motorola Appl. Note. *AN-475*

5. *Gated Video Amplifier Applications, The MC1545*. Motorola Appl. Note *AN-491*

6. *MC 1596 Balanced Modulator*, Motorola Appl. Note, *AN-531*

Exercises 5

5.1 Consider the ways in which multiplier devices and multiplier devices and filters can be used to perform: amplitude modulation and demodulation; supressed carrier double side band modulation and demodulation; suppressed carrier single side band modulation and demodulation; frequency doubling and phase sensitive detection.

5.2 The following component values are used in the circuit of figure 5.12: $R_b = 12\mathrm{k}\Omega$, $R_x = 10\mathrm{k}\Omega$, $R_L = 3.3\mathrm{k}\Omega$, $V_{s+} = +12$ V, $V_{s-} = -8$ V. Find the voltage levels at pins 6 and 9 with zero input signals and offsets balanced.

5.3 A sinusoidal signal of amplitude 2 volts is applied to the X channel of the device in the circuit of question 5.2. If this same signal is simultaneously applied to the Y channel what is the d.c. component of the differential output voltage? (see sections 4.4.2 and 5.3.1).

6. Four Quadrant Linear Multipliers — Practical Considerations and Applications

Four quadrant linear multipliers are variable gain devices with special properties. Multiplication is a fundamental operation in signal processing but high cost and circuit complexity have, in the past, limited the usefulness of a multiplying element. Like the operational amplifier the multiplier was once mainly used in the specialised field of analogue computation, but again, like the op. amp., the availability of a multiplying module in a low cost integrated circuit form has resulted in its use in an ever widening variety of applications. The product ranges of the manufacturers specialising in circuit modules contain a variety of multiplying devices; we restrict our attention to monolithic multipliers employing the variable transconductance circuit technique discussed in chapter 4.

6.1 Practical Multipliers – Departures from Ideal Behaviour

Ideally a four quadrant multiplier provides a single-ended output signal which is determined by a relationship of the form

$$V_O = K\ V_X\ V_Y \tag{6.1}$$

K is the scaling factor of the multiplier, V_X and V_Y are the signal voltages which are applied to the multipliers two input channels. They may be of either polarity, the polarity of the output signal depends upon the polarity of both input signals.

Practical multiplying devices differ in the amount of external circuitry and in the number of external adjustments that they require and after making all adjustments their performance is still not completely ideal. A multipliers' departure from the ideal introduces errors into any signal processing operation in which it is employed. Unfortunately there are no established standards for multiplier specifications and thus, when attempting to compare the products of different manufacturers, it is important to examine in detail the way in which the performance parameters are defined. Data sheets should be critically examined whenever a multiplying device is to be used in a demanding signal processing operation, for the assessment of the possible accuracy in a practical application is by no means straightforward, 'typical' specifications should be viewed with some caution.

The present absence of an agreed set of multiplier performance specifications is not really surprising in view of the different facilities provided by the various devices. In multiplying devices requiring extensive external circuitry, overall performance is obviously not solely a characteristic of the multiplying device but is equally dependent upon the performance parameters of the external circuitry. Monolithic multipliers have been introduced only comparatively recently, the next few years is likely to see the development of many new types and it seems that we must wait for the 'standard' multiplier to emerge before we can expect a standard set of performance parameters. Currently available multipliers are extremely versatile devices which the practising engineer cannot afford to neglect and if he is to select and intelligently use a multiplier he must be prepared to sort out the differences in terminology between the different manufacturers. Monolithic multipliers are not really difficult to use in practice and an experimental evaluation of the behaviour of a particular device does much to clarify the significance of its performance parameters; the reader is urged to perform such a practical evaluation for himself. We present a general discussion of multiplier parameters that it is hoped will give the reader greater confidence when he attempts to interpret device data sheets.

We discuss multiplier parameters in terms of the functional diagram illustrated in figure 6.1., the reader should relate this functional description to the more detailed treatment of multiplier circuit techniques given in section 4.5. Figure 6.1 gives a general functional description; particular multiplier devices may be expected to differ in detail. In some monolithic multipliers the output op. amp., shown in figure 6.1, forms an integral part of the device, in others it must be connected externally. Differences also exist in regard to offset and scale setting adjustments.

The reader will be assumed to have an understanding of op. amp. specifications; many of the parameters used to describe the characteristics of a multiplier are essentially similar to those used in specifying the performance of an op. amp. Differences exist and extra parameters are required by a multiplier because unlike an op. amp., a multiplier has two input channels. Multiplier input channels are differential, differential input voltage range and maximum common mode input voltage specify input limitations. The parameters, input resistance, input bias current, input offset current, input offset voltage, and their temperature dependence are required to further specify the characteristics of a multipliers two input channels. The relative importance of the temperature dependence of input bias currents and input offset voltages in their effect on accuracy is much the same as for an operational amplifier.

Linear Integrated Circuit Applications

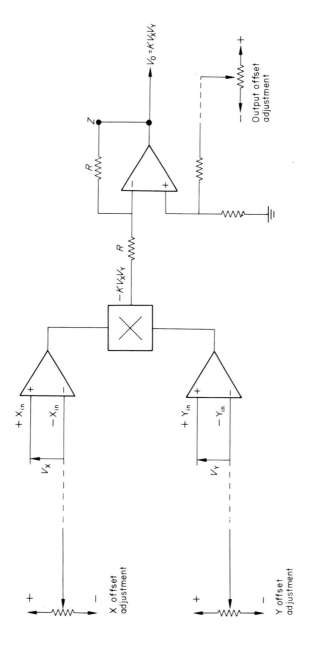

Fig. 6.1 Four quadrant multiplier, general functional schematic

The relationship between multiplier specifications and multiplier errors may be conveniently interpreted in terms of a general expression for the output voltage given by a real multiplier (as distinct from the ideal relationship of equation 6.1), such a relationship may be written as

$$V_O = K(1+\Delta)V_X V_Y + V_{fx} + V_{fY} + K(1+\Delta)V_{iox} V_{ioY} + V_{oo} + f(V_x, V_Y) \tag{6.2}$$

where: V_x is the differential voltage applied to the X input channel
V_Y is the differential voltage applied to the Y input channel
K is the multipliers' scale factor
Δ is the scale factor error
V_{fx} is the 'X' signal feedthrough
V_{fY} is the 'Y' signal feedthrough
V_{oo} is the output offset voltage
$f(V_x, V_Y)$ is a non linearity term
V_{iox} is the input offset voltage of the X input differential amplifier
V_{ioY} is the input offset voltage of the Y input differential amplifier

6.1.1 Feedthrough and Offsets

The ideal multiplier, in accordance with equation 6.1 should give zero output when either of its two inputs are zero, regardless of the signal applied to the other input. In the case of a practical multiplier several factors can contribute to a non-zero output and give rise to errors. The operative factors are represented in equation 6.2 by the feedthrough and offset terms.

V_{fY} represents an output voltage component which varies with the Y input signal and which is present even when the X input signal is made zero. Similarly V_{fx} represents an output voltage component which varies with the X input signal and which is present even when the Y input signal is made zero.

The output offset voltage term, $V_{oo} + K(1+\Delta)V_{iox}V_{ioY}$, represents a d.c. component in the output voltage which is independent of both the X and Y input signals. V_{oo} is associated with the output operational amplifier, it is balanced out by the use of a suitable output offset adjustment potentiometer.

The feedthrough terms are each made up of two components, a linear component which is due to an imperfect balance in the input differential amplifier and a non-linear component which arises because of non-linearity in the multiplying circuit element. We may write

$$V_{fx} = K(1+\Delta)V_x V_{ioY} + v_{fx}$$
$$\phantom{V_{fx} = }\underbrace{\phantom{K(1+\Delta)V_x V_{ioY}}}_{\text{varies linearly with } V_x} \underbrace{\phantom{v_{fx}}}_{\text{varies non linearly with } V_x}$$

$V_{fY} = K(1 + \Delta) V_Y V_{iox} + v_{fY}$

varies linearly   varies non linearly
with V_Y with V_Y (6.3)

V_{iox} and V_Y represent the d.c. input offset voltages of the X and Y differential input amplifiers; their effects can be balanced out by using the X and Y offset adjustment potentiometers. When input offset voltages have been balanced there remains the untrimmable non-linear component of feedthrough. Feedthrough specifications for a multiplier, when they are given, usually refer to the signal fed through when an input is exercised throughout its maximum range (with the other input zero), and when input offsets have been balanced out. Currently available monolithic multipliers give a feedthrough signal of typically 50 mV p-p. when one input signal is made zero and a low frequency sinusoidal signal of 20 V p-p. is applied to the other input. The feedthrough signal is not purely sinusoidal but consists of a d.c. term, a fundamental and harmonics (mainly second harmonic). The feedthrough signal in general shows a marked increase at higher frequencies.

The significance of feedthrough and offsets will be readily appreciated by the reader if he performs an offset adjustment on a practical multiplier. Referring to figure 6.1 an offset balancing procedure would be as follows.

1. Connect the $+X_{in}$ terminal to earth and apply a low frequency sinusoidal signal to the $+Y_{in}$ terminal, (say 5 V p-p. 200 Hz).
2. Adjust the X offset adjustment potentiometer for a minimum a.c. signal output.
3. Interchange the connections to the $+X_{in}$ and $+Y_{in}$ terminals and adjust the Y offset adjustment for a minimum a.c. signal output.
4. Connect the $+X_{in}$ and $+Y_{in}$ terminals to earth and adjust the output offset potentiometer for zero d.c. output. In practice the ability to accurately trim offsets will depend upon the quality of the offset adjustment potentiometers, potentiometers of the 'infinite resolution' variety are to be preferred.

6.1.2. Scale Factor and Scale Factor Errors

In many monolithic multipliers the value of the scale factor is set by external means, a value $K = 1/10$ is normally used. This allows input signal amplitudes $V_X = V_Y = \pm 10$ V with a maximum output voltage of ± 10 V which lies within the output voltage capability of most op. amps.

The circuit parameters which determine the scale factor of a variable transconductance multiplier were analysed in section 4.5 and it was shown that the scale factor is given by the relationship

$$K = \frac{2 R_L}{I R_X R_Y} \qquad (6.4)$$

R_X and R_Y are emitter degeneration resistors associated with the input differential amplifiers and R_L is the resistor used to convert the output current given by the multiplying element into an output voltage, (R_L is the feedback resistor connected to the output op. amp. in figure 6.1). In some devices, for example the Analog Devices AD 530 or Intronics M530, R_X, R_Y, R_L and I are all fixed internally so as to give a value of K somewhat greater than 1/10. The effective value of K is then externally trimmed to be exactly 1/10 by externally attenuating one of the input signals to the multiplier. In devices like the MC 1594/1494 and the MC 1595/1495, R_X, R_Y, R_L are external resistors and the value of I is determined by an external resistor. The values of R_X and R_Y are chosen by the user in accordance with the maximum input signal amplitudes expected in the application and the value of R_L is then trimmed to give the exact value of K required. A requirement for several external resistors represents an added design consideration and an increase in external circuit complexity but it does make for somewhat greater versatility. For example, if input signal amplitudes are considerably less than the allowable input range of ± 10 V, it may be an advantage to use a scale factor $K = 1$ or even $K = 100$ and this can be done when scale-setting components are connected externally.

The scale factor error term, Δ, represents the departure of the actual scale factor from its desired value. The error is trimmed to zero in the scale-setting procedure but it may be expected to change with time, temperature, power supply voltage and with the frequency of the input signals. The only significant scale factor errors, except in the most critical applications, are normally those associated with inadequate bandwidth.

Frequency response parameters are determined by the characteristics of the output op. amp. and as such are specified by small signal bandwidth and slew rate. In general, the small signal frequency response of a multiplier exhibits a 20 dB/decade roll off at frequencies beyond the 3 dB bandwidth governed by the frequency compensation which is applied to the output op. amp.

As would be expected of a 1st order frequency response, amplitude attenuation with increase in frequency is accompanied by a phase shift but in many applications it is the amplitude error alone which is significant. In applications in which the phase of the output signal is important, for example in double sideband modulation and demodulation, phase shift can contribute a significant error at frequencies at which the amplitude error is negligible. The effect is specified in terms of a vector representing the difference between the ideal and

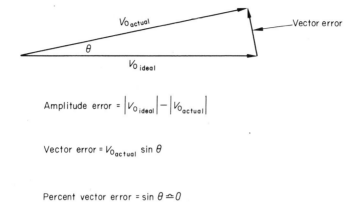

Fig. 6.2   Representation of vector error

actual output signal given by the multiplier which is called the phase vector error, the vector error is illustrated in figure 6.2. Note that a phase shift of 0.57° is equivalent to a phase shift of $0.57 \times 2\pi/360 = 0.01$ radians and gives a 1 per cent vector error, whereas the amplitude error is only 0.005 per cent at this phase shift. In the case of a double sideband modulator or demodulator a 1 per cent vector error represents a 1 per cent amplitude error at the phase angle of interest.

6.1.3. *Multiplier Non-Linearity*

The output voltage given by a practical multiplier is not exactly proportional to the product V_X, V_Y even when feedthrough and offsets are trimmed. Practical multipliers exhibit a non-linear behaviour represented in equation 6.2 by the non-linearity error term $f(V_X, V_Y)$. Non-linearity is normally small for small output voltages but increases as the output voltage approaches its full scale value.

Non-linearity is normally specified in terms of the maximum deviation of the output voltage from a straight line function (expressed as a percentage of the full scale output), obtained when one input signal is fixed at its maximum value whilst the other is exercised through its allowable range. The significance of the specification is illustrated by the multiplier transfer curves shown in figure 6.3. Note that the non-linearity associated with the two input channels is not normally the same and a separate non-linearity specification is required for each channel. The non-linearity of a practical multiplier can, in principle,

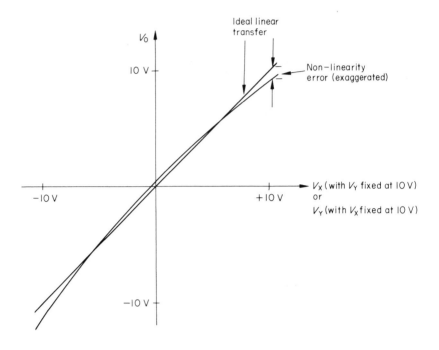

Fig. 6.3   Multiplier transfer curve

be measured from transfer curves of the type shown in figure 6.3 that have been obtained experimentally using an X/Y recorder. In practice small departures from linearity are difficult to measure from a transfer curve and an a.c. test circuit, which will be described later, is normally used for non-linearity measurements.

An estimate for the non-linearity term in equation 6.2 can be made from a knowledge of the non-linearity specifications and a use of the conservative approximation

$$f(V_X, V_Y) = |V_X| \epsilon_X + |V_Y| \epsilon_Y \qquad (6.5)$$

ϵ_X and ϵ_Y are the specified non-linearities for the X and Y inputs respectively expressed as fractions. A numerical example should serve to clarify the significance of equation 6.5.

A certain multiplier has the following non-linearity specifications; X input 0.8 per cent, Y input 0.2 per cent. What is the maximum non-linearity error which can be expected for $V_X = 5$ V, $V_Y = 1$ V?

The expected non-linearity error
$$f(V_X, V_Y) = 5 \times 0.008 + 1 \times 0.002$$
$$= 0.042 \text{ V}.$$
The nominal output
$$\frac{V_X V_Y}{10} = 0.5 \text{ V}$$

The non-linearity error represents an 8.4 per cent output error. Note that if the X and Y inputs are interchanged the non-linearity error becomes
$$1 \times 0.008 + 5 \times 0.002 = 0.018 \text{ V}$$

This represents an output error of 3.4 per cent. If the input signals to a multiplier have different amplitude ranges it is clearly advantageous to apply the signal with the smaller range to the input channel which has the greater non-linearity associated with it.

6.1.4 Total D.C. Error

Multiplier data sheets sometimes give a total d.c. error specification. The error is given as a percentage of full scale output. It represents the maximum deviation of the actual output voltage from that predicted by the ideal multiplier equation, equation 6.1, and it gives an indication of the minimum overall d.c. accuracy obtainable with the multiplier. The total error includes all error factors other than frequency dependent effects, that is, it includes feedthrough, offset, scale factor and non-linearity errors which we have discussed and which are separately specified. The separately specified errors do not necessarily augment one another, in fact some errors tend to cancel one another when they are combined. Total d.c. error is specified after the multiplier has been externally (or internally), trimmed for optimum performance.

A single total d.c. error specification does not, unfortunately, give a complete answer to multiplier accuracy, accuracy varies over the four quadrants. A more complete accuracy specification would require a measurement of the deviation of the output voltage from its predicted value for various combinations of the input voltages.

Total d.c. accuracy varies with temperature and supply voltages. Parameters which indicate the expected variation are given on some device data sheets.

6.2 Multiplier Test Circuits

In order to gain an initial familiarity with monolithic multipliers and their characteristics it is a useful exercise to connect up practical test circuits and use them to investigate multiplier action and to measure values of performance parameters.

174 Linear Integrated Circuit Applications

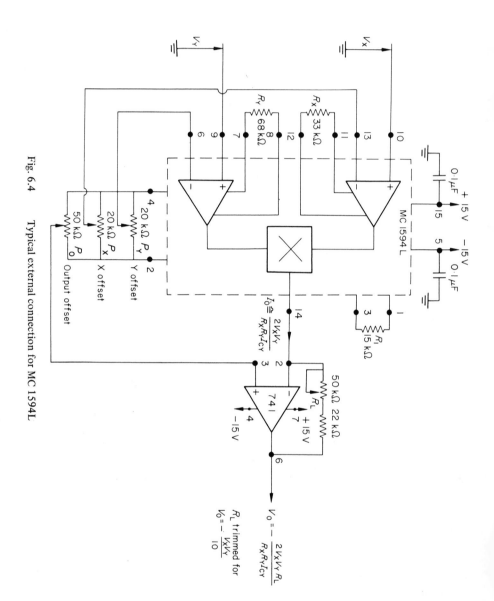

Fig. 6.4 Typical external connection for MC 1594L

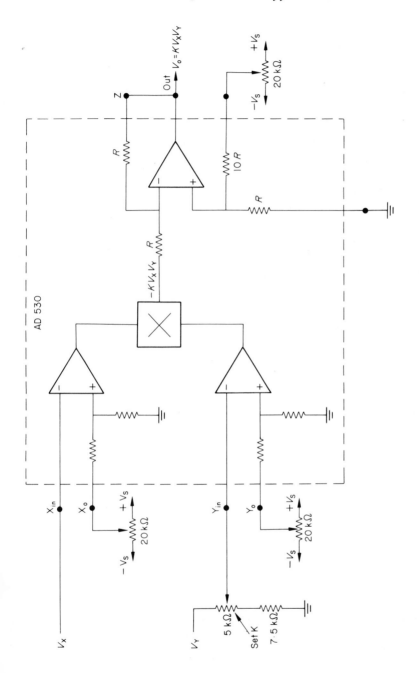

Fig. 6.5 AD530 functional schematic; multiplier connections

6.2.1 Basic Circuit Arrangements

Currently available monolithic multiplying devices have varying external circuit requirements. The most recently introduced i.c. multipliers are internally trimmed prior to final packaging, (e.g. Analog Devices, AD 532, Burr-Brown 4203) and can be used without external offset adjustments. Devices like the AD 530 and the Burr-Brown 4201 have a requirement for four trim potentiometers and one resistor, but the MC 1594/1494 device requires four trim potentiometers, plus an external op. amp., plus additional resistors and capacitors. Suggested connections for multiplier operation of the MC 1594/1494 and the AD 530 are given in figures 6.4 and 6.5 respectively. In each case a functional representation of the internal circuitry of the device is given.

Referring to figure 6.4 resistor R_1 is used to set a value for the constant current sources in the multiplier circuitry (see section 4.5); a value $R_1 = 16\text{k}\Omega$ is recommended by the manufacturers and is said to set the current sources at 0.5 mA (in equation 6.4 $I_{CY} \cong 0.5$ mA). The nearest preferred value to 16kΩ has been used in figure 6.4. Values of R_X and R_Y are selected according to the maximum expected values of V_X and V_Y, (in figure 6.4 values for $|V_{X_{max}}| = |V_{Y_{max}}| = 10$ V are used). The manufacturers recommend that values be chosen so as to satisfy the inequalities

$$R_X \geqslant 3 \, |V_{X_{max}}| \qquad (R_X, R_Y \text{ in k}\Omega$$
$$R_Y \geqslant 6 \, |V_{Y_{max}}| \qquad V_{X_{max}} \quad V_{Y_{max}} \text{ in volts}) \quad (6.5)$$

The MC 1594/1494 device provides regulated voltages at pins 2 and 4, (pin 2, + 4.3 V, pin 4, $-$ 4.3 V approximately); the offset adjustment potentiometers P_X, P_Y and P_O are connected between these device terminals.

The output signal produced by the MC 1594/1494 is in the form of a single-ended current source referenced to earth. A current to voltage conversion is performed by the operational amplifier shown in figure 6.4. The multiplier scale factor is trimmed by adjusting the value of the feedback resistor which is connected to the amplifier. The frequency response characteristics of the multiplier circuit are largely determined by the bandwidth and slewing rate of the particular amplifier used in the circuit and in applications in which only the a.c. component of the output signal is important the operational amplifier can be omitted from the circuit. The output voltage is then developed across a load resistor connected between the device output terminal and earth and the bandwidth is determined by the time constant associated with this resistor and the stray capacitance at the output terminal. A second frequency dependent factor arises because of an effective bypassing of resistors R_X and R_Y which occurs at the higher frequencies and which leads to an apparent increase in scale factor (gain), at these frequencies. The values of resistors R_X, R_Y and R_L should be kept as small as possible for wide bandwidth operation.

The AD 530 device, as shown in figure 6.5, requires fewer external components than the MC 1594/1494. In the AD 530 the output op. amp. and the main scale-setting components are all included inside the device, the scale factor is usually externally trimmed to be exactly 1/10. In general the accuracy of multiplying devices is greatest when both inputs and output swing through their full range and if either of the input signals to the AD 530 is significantly less than the full range input it may be necessary to use an external op. amp. at the input to provide an appropriate amplification. If such an amplifier is used, its output slew rate must be adequate for the frequency of the signals which are applied to it.

The MC 1594/1494 device can be modified for use with input signals smaller than 10 V by simply using a smaller value of emitter degeneration resistor (R_X or R_Y), and the multiplier scale factor is then easily adjusted to a value which will ensure a full range output signal.

Offsets and scale factor must be trimmed before using a multiplier. Data sheets recommend slightly different trim procedures for the two devices. A procedure for trimming offsets is outlined in section 6.1.1; scale factor is trimmed after completing the offset adjustments. The scale factor may be trimmed to a value 1/10 by applying a 10 V d.c. signal to each input and adjusting the scale setting potentiometer in order to obtain a 10 V output signal.

6.2.2 *Measurement of Input Offsets and Bias Currents*

Input Offset Voltages With input offsets trimmed the voltages at the wipers of the X and Y, offset potentiometers give values for V_{ioX} and V_{ioY} directly. Measurements should be made with a high impedance d.c. millivoltmeter.

Input Bias Currents Multiplier input terminals require bias currents and a d.c. path must always be provided for these currents. An estimate of the bias current taken at a particular input terminal can be made by adopting the following procedure.

Earth the Y input to the multiplier, set V_X at 10 V and read the value of the output voltage. Connect a 1MΩ resistor between the Y input terminal and earth and read the new value of the output voltage. Calculate the value of the bias current I_B from the relationship

$$V_O = K\ V_X\ (-I_b\ 10^6)$$
$$= 1/10\ \ 10(-I_b\ 10^6)$$

A similar procedure may be used to determine the bias currents drawn by the other input terminals.

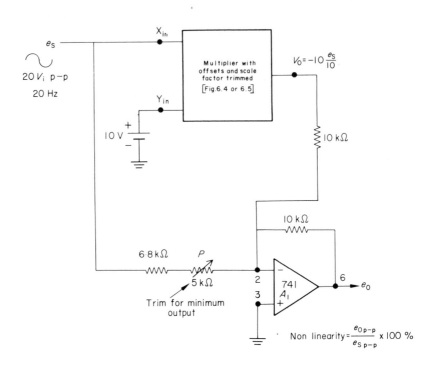

Fig. 6.6   Linearity test circuit

6.2.3 Measurement of Non-Linearity

An a.c. test circuit suitable for measuring the non-linearity associated with the input signals to a multiplier is shown in figure 6.6. A 10 V d.c. signal is applied to the Y input of the multiplier and a low frequency, $(f \cong 20$ Hz$)$, 20 V p-p. sinusoidal signal is applied to the X input. The multiplier output signal is applied together with the alternating X input signal to amplifier A_1 which is connected as an inverting adder. The polarity of the d.c. input signal to the multiplier is such that the multiplier produces a phase inversion of the a.c. input signal and the two signals applied to A_1 subtract. The trimming potentiometer P is adjusted for a null in the output of amplifier A_1 and the output signal of A_1 at null is then due to the distortion components in the multiplier output which arise because of multiplier non-linearity.

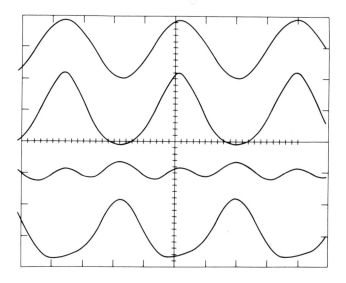

Fig. 6.7   Non-linearity test; typical waveforms

Typical waveforms obtained in the test are shown in figure 6.7; the upper trace (at 10 V/div), represents the X input signal, the other three traces (at 0.2 V/div), show the output signal of A_1 as the potentiometer is adjusted through the null. The peak to peak output of A_1 at null is approximately 0.12 V, this represents a multiplier non-linearity of $0.12/20 \times 100\% = 0.6$ per cent.

6.2.4 *Frequency Response Characteristics*

Bandwidth and output slewing rate characteristics of a multiplier are measured by applying a d.c. reference signal to one input channel and a varying test signal to the other input channel. For example, in a measurement of small signal bandwidth a 10 V d.c. signal is applied to one channel and a small, constant amplitude, sinusoidal signal is applied to the other. The amplitude of the multiplier output signal is measured as the frequency of the alternating input signal is varied and the small signal frequency response is then plotted as a dB/log f graph. In the case of slewing rate measurements the sinusoidal test signal is replaced by a low frequency square wave signal of amplitude 10 V and measurements are made on the multiplier output waveform which is monitored by an oscilloscope.

6.3 Multiplier Applications

The capabilities of monolithic four quadrant multipliers in signal processing applications are as yet not fully realised, they remain to be developed by the ingenious systems designer. A selection from the wide range of existing applications is presented in the hope that they will encourage the reader to further explore the ways in which these most versatile devices can be applied. The first application example is presented here in terms of an externally trimmed multiplier but for the sake of simplicity most of the other applications are shown with an internally trimmed device used in the circuits. Applications can in fact generally be implemented with any multiplier device, the decision as to which device to use must ultimately be made by the systems designer himself on a cost performance basis. External component requirements and the time taken to make any necessary trimming adjustments should not be overlooked when assessing costs.

6.3.1. *Squaring, Dividing, Square Rooting*

The basic multiplier circuit arrangements of figures 6.4 and 6.5 can both, with very minor modifications, be used to perform squaring, dividing and square rooting operations and an internally trimmed multiplying device like the AD 532 is even more convenient for these operations.

Squaring

If the X and Y input terminals in the circuits of figure 6.4 and figure 6.5 are connected together the circuits perform a squaring operation on any signal applied to this common input terminal. In a multiplier operation the input offset voltages associated with the X and Y differential input channels must be separately balanced. A consideration of the multiplier performance equation, (equation 6.2), shows that only a single input offset adjustment is strictly necessary for a squaring operation. If we write $V_X = V_Y = V_i$ and neglect untrimmable components of feedthrough and non-linearity equation 6.2 becomes

$$V_O = K(1 + \Delta) V_i^2 + K(1 + \Delta) V_i (V_{iox} + V_{ioY}) + V_{oo} \quad (6.7)$$

Equation 6.7 shows that in a squaring operation, input offsets are represented by a composite offset term which, because it is the algebraic sum of V_{iox} and

V_{ioY} can be nulled by a single offset adjustment, say the X offset. The Y offset potentiometer can be omitted from the circuit and the unused Y offset input terminal is connected to earth. The X offset potentiometer is, in effect, adjusted to leave a resultant X offset which is equal and opposite to V_{ioY} and which makes the second term in equation 6.7 zero.

A setting up procedure for the squaring circuit is as follows

1. Make $V_I = 0$ V. and adjust the output offset potentiometer for zero d.c. output (makes $V_{oo} = 0$)
2. Make $V_i = 10$ V d.c. and adjust the scale factor trim for 10 V output (-10 V for MC 1594/1494, $+10$ V for AD 530 circuit)
3. Reverse the polarity of V_i (this reverses the polarity of the second term in equation 6.7), the output voltage becomes, say, $10 + e$. Adjust the X offset potentiometer to make the output voltage $10 + e/2$ (this zeros the second term in equation 6.7). Re-adjust the scale factor trim to make the output voltage 10 V.
4. Check the output offset with $V_i = 0$ and if non zero repeat the above procedure until no errors remain.

A squaring circuit can be used for frequency doubling, (see section 5.3.1., equation 5.3), the action is illustrated by the waveforms in figure 6.8. The lower

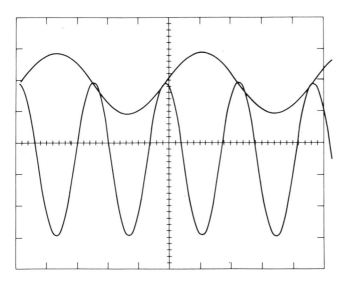

Fig. 6.8 Squaring circuit used for frequency doubling; input sinusoid 5V/division, output second harmonic 0.5V/division. Both traces have same vertical zero reference

trace shows an input sinusoid (at 5 V/div), and the upper trace shows the output waveform of the squaring circuit (at 0.5 v/div), both waveforms have the same vertical zero reference.

Dividing and Square-Rooting

In the dividing and square-rooting mode of operation of a multiplier the multiplying section of the circuit is used as a feedback element for the operational amplifier, the necessary circuit connections are illustrated in figures 6.9 and 6.10. The applications are illustrated with reference to the internally trimmed multiplier device, type AD 532. An internally trimmed device gives the user the convenience of a minimum requirement for external components and adjustments. Other multiplying devices can be used for these applications and the reader is referred to the appropriate data sheets for details of the trim procedure required by externally trimmed devices when they are used for division and square rooting.

Referring to figure 6.9, the output signal from the op. amp. is fed back to its input via the Y channel of the multiplier. The feedback fraction and the closed loop gain of the amplifier are controlled by the signal applied to the X channel of the multiplier. The sign of the X signal must be such that there is no phase inversion of the signal fed back through the multiplier, otherwise feedback becomes positive and the amplifier saturates. The X signal is thus restricted to negative values; positive values of V_X can be manipulated by connecting them to the $-X_{in}$ terminal and earthing the $+X_{in}$ terminal.

If the operational amplifier is assumed to behave ideally the usual summing point restraints give

$$-\frac{V_X V_Y}{10 R} + \frac{V_Z}{R} = 0$$

but $V_Y = V_0$

thus $V_0 = 10 \dfrac{V_Z}{V_X} = -\dfrac{10 V_Z}{|V_X|}$ [Note V_X must be negative]

As the magnitude of V_X is decreased towards zero the closed loop gain of the op. amp. approaches its open loop value producing a proportional degradation in accuracy, noise and frequency response. If ϵ_m is the total error specification for the device used in the multiplier mode the output error when used in the divide mode is approximately 10 ϵ_m/V_X. Note that a total output error of 1 per cent for $V_X = 10$ V becomes 10 per cent if V_X is reduced to 1 V. The circuit is not suitable for use with small values of V_X since small values of the denominator always give rise to

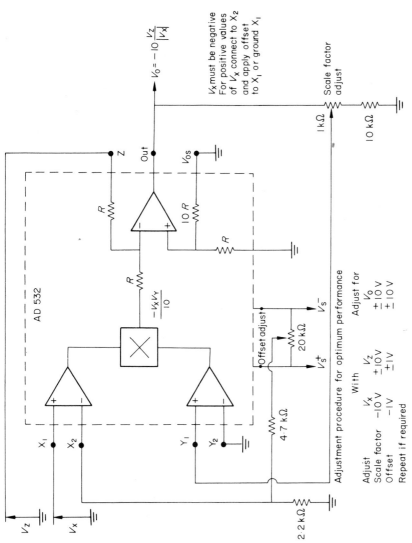

Fig. 6.9　Multiplier used for division

184 Linear Integrated Circuit Applications

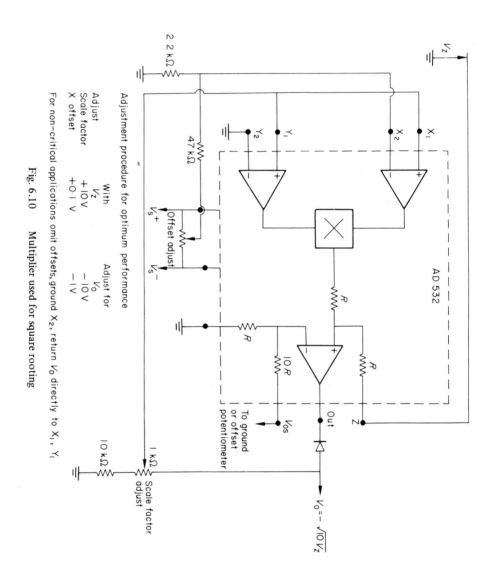

Adjustment procedure for optimum performance

Adjust	With	Adjust for
Scale factor	V_z	V_0
	+10 V	−10 V
X offset	+0.1 V	−1 V

For non-critical applications omit offsets, ground X_2, return V_0 directly to X_1, Y_1

Fig. 6.10 Multiplier used for square rooting

large errors when a multiplier is connected in the divide mode. A log divider should be considered for applications requiring accuracy over a wide dynamic range of signal levels.

The square rooting mode of operation for a multiplier is illustrated in figure 6.10; feedback to the operational amplifier in the device is again supplied via the multiplying circuit element, but in this case the amplifier output voltage drives both the input channels of the multiplier. The signal fed back to the amplifier summing point thus depends on V_o^2. In order that feedback shall be negative only negative values of output voltages are allowed and operation is restricted to positive values of V_Z. The feedback forces the relationship,

$$V_0 = -(10\,V_Z)^{1/2}$$

The diode D is included in the circuit to prevent possible latch-up which might otherwise occur for very small values of V_Z. The output offset trim, if it is included in the circuit, should be adjusted to make $V_0 = -1$ V with $V_Z = 1$ V. Note that negative values of V_Z can be manipulated to give positive values of the output voltage by interchanging the connections to the $-X_{in}$ and $+X_{in}$ terminals and reversing the polarity of the diode.

6.3.2 *Mean Square and Root Mean Square*

The mean square and root mean square values of alternating signals can be readily measured using four quadrant linear multipliers. A multiplier, connected in the squaring mode, produces an output signal which is proportional to the square of the instantaneous input signal applied to it. If the squarer is followed by an averaging filter, (a low pass filter), the output from the filter represents the mean square value of the input signal. A second multiplier, connected in the square rooting mode, can then be used to form the root mean square value.

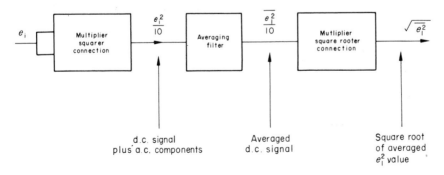

Fig. 6.11   Multiplier used to give RMS value

The system is shown in figure 6.11.

A multiplying device with a dynamically adjustable scale factor, (e.g. the AD 531, see section 4.5, figure 4.12), allows R.M.S. measurements to be made using a single multiplying device, a circuit for this purpose is illustrated in figure 6.12. The AD 531 device gives an output signal which is determined by a relationship of the form

$$V_0 = \frac{V_X V_Y}{k I_Z}$$

where the current I_Z may be varied by a voltage V_Z which is applied to an operational amplifier used as a voltage to current converter. In figure 6.12 the output signal of the AD 531 is averaged by a simple filter with the output from the filter buffered by a high input impedance operational amplifier follower. The X and Y input channels of the multiplier are connected together making $V_X = V_Y = V_{in}$. The averaged output signal of the multiplier, $V_0 = \overline{V_{in}^2}/V_Z$ is fed back to provide the signal V_Z. Thus $V_0 = V_Z$ and the loop forces the relationship

$$V_0 = \frac{\overline{V_{in}^2}}{V_0}$$

or $V_0 = (\overline{V_{in}^2})^{\frac{1}{2}}$

The reader is referred to the AD 531 data sheet for recommended trim procedures.

The use of a four quadrant multiplier to measure the R.M.S. value of an alternating signal has several advantages over other methods. Many 'R.M.S.' meters simply measure the average value of the rectified wave and the meter is scaled so as to indicate the R.M.S. value on the assumption that the waveform is sinusoidal.

A sinusoidal signal, $v = V \sin \omega t$, has a full wave rectified average value

$$V_{av} = \frac{\int_0^{\frac{\pi}{\omega}} V \sin \omega t \, dt}{\frac{\pi}{\omega}} = 2 \frac{V}{\pi}$$

Its R.M.S. value is

$$V_{RMS} = \sqrt{\frac{\int_0^{\frac{2\pi}{\omega}} v^2 \sin \omega t \, dt}{\frac{2\pi}{\omega}}} = \frac{V}{\sqrt{2}}$$

Linear Integrated Circuit Applications

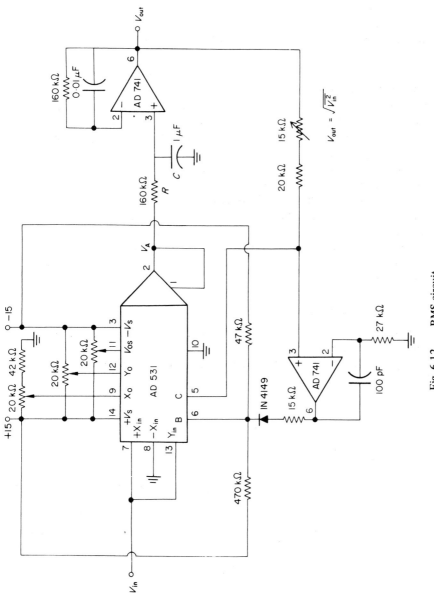

Fig. 6.12 RMS circuit

Thus, for a sinusoidal signal

$$\frac{V_{RMS}}{V_{av}} = \frac{\pi}{2\sqrt{2}} = 1.11$$

and an average reading is converted to R.M.S. by using an indicating scaling factor of 1.11.

This method of measuring R.M.S. values is valid for undistorted sine waves but the R.M.S. indication will clearly be in error if the waveform is appreciably distorted. If the waveform is regular, say triangular or square, an appropriate adjustment of scaling factor allows R.M.S. values to be determined. No such adjustment is possible if the signal to be measured has an arbitrary waveform.

There are instruments which give a true indication of R.M.S. value regardless of waveform, for example, hot wire and thermocouple types but whilst these instruments give a visual indication of R.M.S. value they do not provide an electrical output which is suitable for remote recording or control purposes. A measurement of R.M.S. value with a multiplier gives a true indication regardless of the wave shape since the instantaneous input signal values are squared and averaged. The multiplier gives an output which may be used to drive a visual R.M.S. indicator, in addition the output may be further processed or used for control purposes.

6.3.3 *Power Measurement*

A multiplier may be used to give a direct indication of the amount of power dissipated by a complex load. Circuitry of the type shown in figure 6.13 is used, resistance R is a current sensing resistor connected in series with the load.

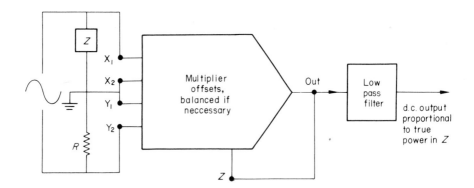

Fig. 6.13   Power measurement

Linear Integrated Circuit Applications

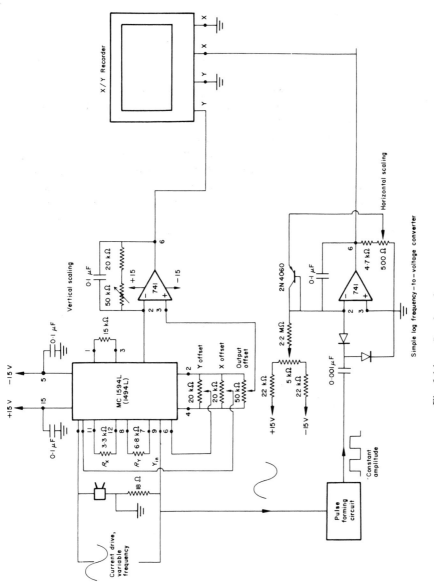

Fig. 6.14 Loudspeaker power measurement

If the circuit is supplied with an alternating current $I \cos \omega t$ the multiplier output voltage has the form

$$V_0 = \frac{I Z (\cos \omega t + \theta) I R \cos \omega t}{10}$$

or $\quad V_0 = \dfrac{R}{20} I^2 Z \cos \theta + \dfrac{R}{20} I^2 Z \cos 2\omega t$

The first term is a d.c. term which is directly proportional to the true power $I^2 Z \cos \theta$, the second term is an a.c. term which may be readily filtered out using a low pass filter. If the multiplier input signals $(X_1 - X_2)$ and $(Y_1 - Y_2)$ are considerably less than the permitted input swing ($\pm$ 10 volts) the signals should be amplified before applying them to the multiplier.

As an alternative to pre-amplification of multiplier input signals a multiplier type which provides gain (scale factor K greater than 1 rather than the usual $K = 1/10$th) can be used. The circuit shown in figure 6.14 represents an example of the use of this type of device. It is a circuit used to measure the power consumed by a small loudspeaker at different frequencies. A simple log of

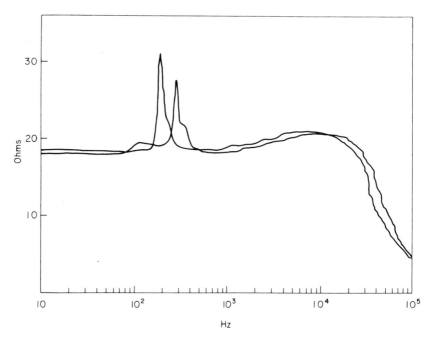

Fig. 6.15   Loudspeaker power

frequency to voltage converter is used to allow a power/log frequency plot to be obtained. The vertical power scale was calibrated in terms of equivalent resistance by substituting known standard resistors for the loudspeaker. Typical X/Y plots obtained with the system are shown in figure 6.15. Frequency was swept manually, the two plots shown were obtained, one with the loudspeaker free standing and the other with the loudspeaker in an infinite baffle type of enclosure.

6.3.4 Automatic Level Control Applications

Four quadrant multipliers readily lend themselves to voltage amplitude stabilisation and other level control applications. The principle is illustrated in the voltage stabilising circuit shown in figure 6.16[1]. The circuit operation is as follows; the rectified output signal is compared with a d.c. reference signal and the resulting difference is filtered by an integrator. The integrator's output is applied as a control signal V_X to a multiplier connected in the divide mode, the output signal of the multiplier conforms to the relationship:

$$V_0 = \frac{-10 \, V_Z}{V_X}$$

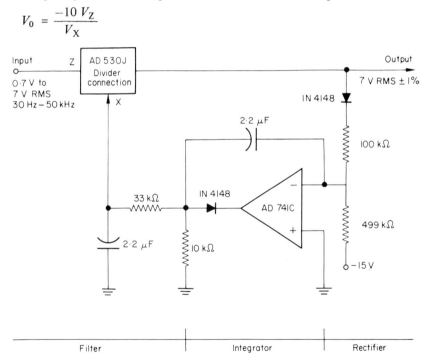

Fig. 6.16   Automatic level control

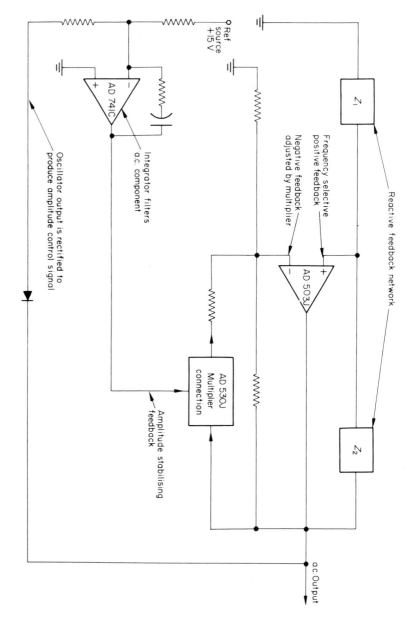

Fig. 6.17 Multiplier in AGC circuit stabilises oscillator amplitude

The larger V_X the smaller is the signal transfer through the multiplier, thus any change in the input signal V_Z develops a value of the control signal V_X necessary to compensate for the input change and hold the output amplitude constant.

Oscillator Stabilisation

The oscillator circuit of figure 6.17 shows a practical application for an automatic level control configuration. In this instance, the multiplier controls the amplitude of the oscillations developed by a Wien bridge oscillator.

Although the principle is much the same as the amplitude stabilising circuit of figure 6.16, one difference lies in the multiplier rather than the divider connection. This is because the resistance capacitance oscillator of figure 6.17 uses the multiplier to control the amount of negative feedback applied to the amplifier inverting input terminal. As the output signal increases from its desired amplitude, the multiplier output increases, causing an increase in the negative feedback to the amplifier input terminal and forcing the output signal amplitude back to its correct level. The rectified difference between output

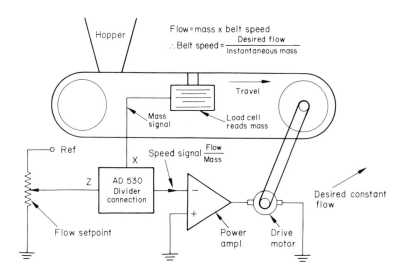

Fig. 6.18   Conveyor flow control

signal and reference is applied by the integrator so as to pass more (not less) signal through the multiplier, as the oscillator output rises. Conversely, reduced oscillator output cuts back the multiplier control signal, thereby lowering negative feedback and returning the circuit's output towards its correct level.

Conveyor Flow Control 1

The principle of using a multiplier to hold some parameter at a pre-determined level is further demonstrated by the application outlined in figure 6.18. Particle flow is the variable being stabilised in figure 6.18, the arrangement adjusts motor speed, hence velocity of the conveyor belt, to deliver a constant rate of flow. The stabilising circuit is necessary because of variations in the amount of material on the conveyor belt. The load cell monitors the instantaneous mass of material on a section of the conveyor belt and this signal is used to control the motor speed.

6.3.5 *Voltage Controlled Quadrature Oscillator* 1

The circuit shown in figure 6.19 illustrates a method of using two multipliers to allow a voltage control of the frequency of the two phase oscillator. The circuit in figure 6.19 is based on a pair of cascaded integrators, each of which contributes $90°$ phase shift at the operating frequency, plus a unity gain inverter circuit which provides the final $180°$ phase shift necessary to fulfil the conditions for oscillation. Oscillator frequency is adjusted simply by altering the time constants of the two integrators. In the circuit shown, a 1 to 10 volt control signal range produces a 100 Hz to a 1000 Hz frequency swing. Other component values could, of course, place this 10:1 frequency range at different levels within the multipliers operating band.

6.3.6 *Further Computation Circuits*

Vector Sums and Differences The need sometimes arises for computing circuits which give an output signal related to two input signal voltages by an equation of the form

$$(V_1^2 + V_2^2)^{1/2} \text{ or } (V_1^2 - V_2^2)^{1/2}$$

A circuit configuration for a direct computation would involve the use of two multipliers (as squarers), an operational amplifier for summation (or differencing) followed by a third square root connected multiplier. The direct approach is likely to be expensive and indirect methods of computation can often reduce the number of multipliers required.

A single four quadrant multiplier with a differential input capability can be

Linear Integrated Circuit Applications 195

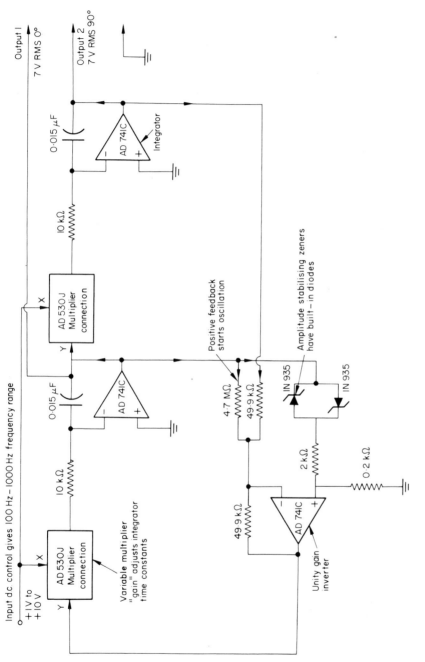

Fig. 6.19　Voltage controlled two-phase oscillator

used to form an output signal proportional to the difference of squares; a circuit for this purpose is shown in figure 6.20. The differential input multiplier gives an output signal

$$V_0 = \frac{(X_1 - X_2)(Y_1 - Y_2)}{10}$$

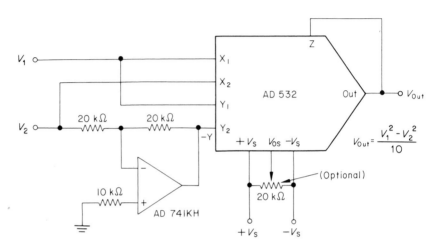

Fig. 6.20   Difference of squares connection

In figure 6.20 $X_1 = V_1$, $X_2 = V_2$, $Y_1 = V_1$
and the unit gain inverter makes $Y_2 = -V_2$

Thus $V_0 = \dfrac{(V_1 - V_2)(V_1 + V_2)}{10}$

or $V_0 = \dfrac{V_1^2 - V_2^2}{10}$

A second multiplier connected in the square root mode allows the computation to be completed.

The use of a multiplying device such as the AD 531 which gives an output signal of the form

$$V_0 = \frac{I_x V_y}{k I_z}$$

allows a vector computation to be performed with a single multiplier; the

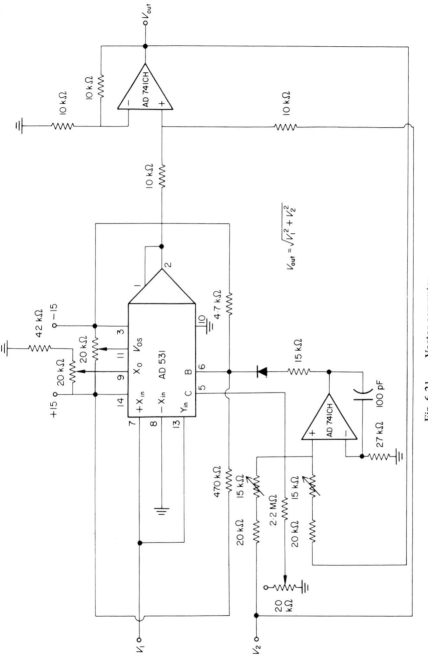

Fig. 6.21 Vector computer

198   Linear Integrated Circuit Applications

principle is illustrated by the circuit shown in figure 6.21. In this circuit the output of the multiplier V_1^2/kI_z is summed with V_2 in the output operational amplifier which gives an output signal $V_0 = \dfrac{V_1^2}{kI_z} + V_2$

The operational amplifier used to supply the current I_z forces the relationship

$$k I_z = V_0 + V_2$$

The complete system thus forces the relationship

$$V_0 = \frac{V_1^2}{V_0 + V_2} + V_2$$

and $V_0^2 = V_1^2 + V_2^2$

$V_0 = (V_1^2 + V_2^2)^{1/2}$

Readers are referred to the AD 531 data sheet for circuit adjustment procedures.

If, in figure 6.21, the $+V_{in}$ terminal is grounded and V_1 connected in the $-X_{in}$ terminal, the circuit forces the identity

$$V_0 = (V_2^2 - V_1^2)^{1/2}$$

Generation of Functional Relationships/Power Series

Arbitary functions can be synthesised by using an operational amplifer, biased diodes and resistive networks. Multipliers can often be used to advantage in non-linear processing applications by eliminating

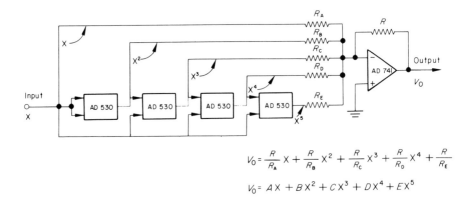

$$V_0 = \frac{R}{R_A}X + \frac{R}{R_B}X^2 + \frac{R}{R_C}X^3 + \frac{R}{R_D}X^4 + \frac{R}{R_E}$$

$$V_0 = AX + BX^2 + CX^3 + DX^4 + EX^5$$

Fig. 6.22 · Function generator

the requirement for diode resistor networks. For example, a square law relationship may be obtained with just one multiplier, whereas a straight line approximated square law response may require as many as a dozen diode resistor networks if the circuit is to be capable of handling dual polarity signals.

All continuous functions can be represented by power series of the form

$$A + BX + CX^2 + DX^3 + \ldots$$

Such a series is readily generated using an operational amplifier together with cascaded multipliers, the circuit principles are illustrated in figure 6.22. Negative coefficients can be readily obtained if a multiplying device with a differential input capability is used.

The generation of a sine function is one application of a power series circuit. A sine wave can be represented by the power series

$$\sin X = X - \frac{X^3}{3!} + \frac{X^5}{5!} - \frac{X^7}{7!} + \ldots$$

The use of an operational amplifier and two multipliers enables a sine wave with little more than 1 per cent distortion to be formed from a triangular wave. The principle is illustrated by the circuit shown in figure 6.23. A linearly increasing voltage applied to the circuit develops an output having a rising sinusoidal shape. Conversely, the mirror image of the rising sinusoid is generated when the linear input decreases from maximum to zero. By increasing the triangular waveform in the reverse direction, then bringing it back to zero again a full cycle of sinusoidal output is produced.

Functional relationships are generated electrically by representing the variables in the functional equation by corresponding voltages in an electrical analogue circuit. The scaling factors used to relate the electrical variables to the functional variables must be chosen by the system designer. Scaling factors used must ensure that the output limitations of the devices in the electrical system are not exceeded. However, for greatest accuracy, voltage ranges should be close to full scale range.

Suitable values for electrical scaling factors can be arrived at by expressing both the functional equation and the performance equation of the analogue electrical circuit in dimensionless form^2. The dimensionless equations may be obtained by multiplying and dividing each variable by its full range value. As an example consider the function

$$y = x - \frac{x^3}{6} \qquad (6.8)$$

with a full range value of x, $x_m = \dfrac{\pi}{2}$

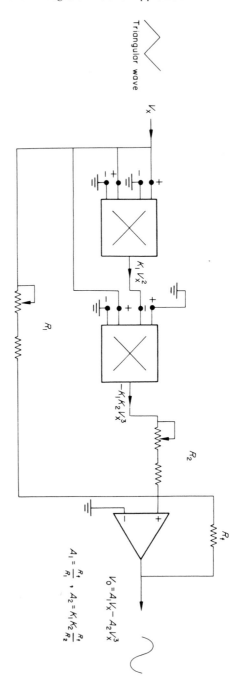

Fig. 6.23 Sine shaping using multipliers

The corresponding full range value of y is $y_m = 0.925$
(7.5 per cent error if used as a single quadrant approximation for a sinusoidal, but see later).

The electrical analogue has the performance equation

$$V_o = V_y = A_1 V_x - A_2 V_x^3 \qquad (6.9)$$

Equation 6.8 may be written as

$$y_m \frac{y}{y_m} = X_m \frac{X}{X_m} - \frac{X_m^3}{6} \left(\frac{X}{X_m}\right)^3$$

or $\left(\dfrac{y}{y_m}\right) = \dfrac{X_m}{y_m} \left(\dfrac{X}{X_m}\right) - \dfrac{X_m^3}{6 y_m} \left(\dfrac{X}{X_m}\right)^3 \qquad (6.10)$

Similarly equation 6.9 may be written as

$$\left(\frac{V_y}{V_{ym}}\right) = A_1 \frac{V_{xm}}{V_{ym}} \left(\frac{V_x}{V_{xm}}\right) - A_2 \frac{V_{xm}^3}{V_{ym}} \left(\frac{V_x}{V_{xm}}\right)^3 \qquad (6.11)$$

Where V_{ym} and V_{xm} represent expected maximum values.

The normalised equations, equations 6.10 and 6.11 must be identical if the electrical analogue circuit is to correctly represent the desired functional relationship, the value of the scaling factors must thus be

$$A_1 = \frac{V_{ym}}{y_m} \frac{X_m}{V_{xm}} \quad : \quad A_2 = \frac{V_{ym}}{6 y_m} \left(\frac{X_m}{V_{xm}}\right)^3$$

The discussion is now related to the practical circuit realisation shown in figure 6.25. If we assume the multiplier scaling factors are $K_1 = K_2 = 1/10$, a value $V_{XM} = 10$ volts will ensure that multiplier output limitations are not exceeded and a value $V_{YM} = 10$ volts will lie within the output capability of the operational amplifier. Suitable values for the scaling factors are thus

$$A_1 = \frac{10}{10} \frac{\frac{\pi}{2}}{0.925} \cong 1.7$$

and

$$A_2 = \frac{10}{10^3} \frac{\left(\frac{\pi}{2}\right)^3}{6 \times 0.925} \cong 0.7 \times 10^{-2}$$

These values of A_1 A_2 may be set by appropriate choice of the scaling resistors R_1 and R_2 used in figure 6.23 and the circuit then functions as an electrical analogue for the generation of the functional relationship

$$Y = X - \frac{X^3}{3!}$$

If only two terms are used as an approximation for a sinusoid a better approximation can be obtained by slightly modifying the coefficients in the equation. A two term approximation with suitably modified coefficients may be used for a single quadrant approximation with only 1.35 per cent error. In the circuit of figure 6.23 the modified coefficients are most conveniently obtained by experimentally trimming scaling factors in order to obtain a sinusoidal output waveform with minimum distortion. Note, by using non-integral exponents, in addition to modified coefficients, a one quadrant two term approximation of sine can be obtained with less than 0.25 per cent error[3].

$$\operatorname{Sin} X = X - \frac{X^{2.837}}{6.182}$$

$$\left(0 - \frac{\pi}{2}\right)$$

Non-integral exponents can be implemented using log techniques.

6.3.7 Modulator/Demodulator Applications

Four quadrant multipliers readily lend themselves to modulator/demodulator applications at frequencies within their bandwidth limitations. The various modulation processes have already been discussed in section 5.3, the use of a variable transconductance four quadrant multiplier rather than a device specifically designated as a modulator, is indicated for applications in which a strictly linear modulation is required. A linear multiplication of two sinusoidal frequencies gives rise to difference frequency components only, whereas a non-linear multiplication produces perhaps an undesired harmonic content in the output product.

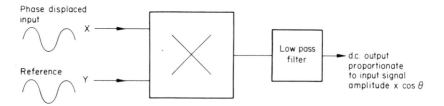

Fig. 6.24   Multiplier used as phase sensitive demodulator

Phase Sensitive Demodulation As an example, we consider a multiplier used to perform the function of a phase sensitive demodulator, figure 6.24. The arrangement is considerably more convenient than many of the circuits commonly used for phase sensitive modulation, both the input and reference signals may be ground referred and no transformers are required. The use of a sinusoidal reference signal means that the output will have no unwanted d.c. components proportional to any odd harmonics in the input signal.

Filtering by Frequency Changing The principle of changing the frequency of a signal before filtering is by no means new and has been used for many years in superheterodyne radio receivers. In these receivers the incoming signal is converted to a fixed frequency, (the I.F.) by modulation using a variable frequency local oscillator, a filter tuned to the I.F. is then used to give the receiver its desired selectivity. This technique overcomes the difficulty of building a variable tuned filter with high Q in the R.F. stages of the receiver. Integrated circuit multipliers and modulators provide a particularly convenient method of frequency changing in filtering applications. The chapter is concluded by a suggestion for an experimental investigation of filtering by frequency changing in connection with a simple spectrum analyser system. The system in no way represents a fully developed spectrum analyser but serves as an experimental demonstration of the principles of spectrum analysis and at the same time it brings together some of the circuits discussed in earlier chapters.

The system, which is shown in figure 6.25, consists of two modulators, two band pass filters, a fixed frequency signal generator, a swept frequency generator and a precise rectifier. The first modulator generates the product of the applied input signal, frequency f_{in}, and the signal from the swept frequency generator, frequency f_1, its output contains frequency components at the sum and difference frequencies $f_1 \pm f_{in}$. The first bandpass filter, centre frequency f_{c1}, passes one of these frequency components ($f_{c1} = f_1 + f_{in}$ in the system shown). The output from the first bandpass filter is mixed with a fixed frequency f_2 in the second modulator and the resultant signal is applied to the second bandpass filter. The second bandpass filter has centre frequency $f_{c2} = f_{c1} - f_2$. A variable Q configuration is used to allow control of the selectivity of the system. The output from the second bandpass filter is rectified by a precise rectifier (peak reading) and applied as the Y deflecting signal for an X/Y plot.

For the purpose of experimental investigation the bandpass filter centre frequencies are not critical. Frequencies used in the system were $f_{c1} = 40\text{kHz}$, $f_{c2} = 8\text{kHz}$ and the frequency f_2 was tuned on test so that f_2 was exactly equal

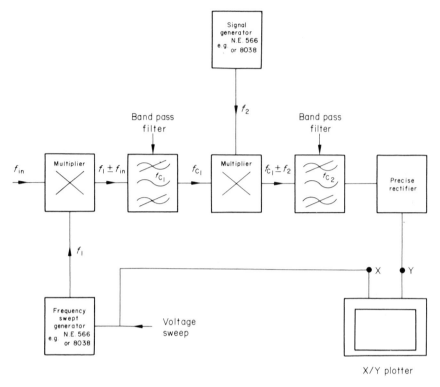

Fig. 6.25   Filtering by frequency changing; simple spectrum analyser system

to $f_{c1} - f_{c2}$. The system gives unambigious results for input frequencies within the range 0 to $2f_{c2}$, i.e. 0 – 16 kHz). Input frequencies outside this range give rise to a spurious response as shown in the experimentally obtained spectra in figure 6.26. In order to obtain a spectra the frequency of the variable frequency generator f_1 was swept from a frequency slightly higher than f_{c1} down to a frequency 18 kHz less than f_{c1}. The first modulator is adjusted to have a small d.c. offset at its f_{in} input channel so that its output always contains a small component at frequency f_1. This provides the spectrum with a zero frequency marker as f_1 passes through the value f_{c1}.

Examination of figure 6.26 shows that when the input signal has a frequency, say $2f_{c2} + f'$ a spurious response appears at $f'|$. At the frequency corresponding to f^1 on the display the swept oscillator has a frequency $f_{c1} - f'$ and the first

Linear Integrated Circuit Applications

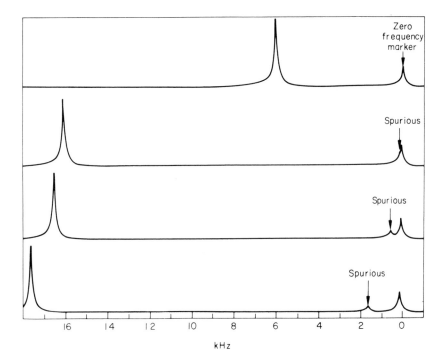

Fig. 6.26   X/Y plots obtained with spectrum analyser system

modulator gives rise to a signal component with frequency $2f_{c2} + f' + (f_{c1} - f')$ $= 2f_{c2} + f_{c1}$ which is not completely rejected by the first bandpass filter. This component gives rise to a difference frequency component in the output of the second modulator at frequency f_{c2} which is passed by the second bandpass filter in the same way as the desired response.

The spectra in figure 6.27 were obtained for an input signal in the form of a square wave of fixed frequency 2 kHz. Note that the upper plot shows several spurious responses due to input harmonics at frequencies higher than 16kHz. The lower plot was obtained using a low pass filter with cut-off frequency 16kHz connected in the signal path before the first modulator, the spurious responses are no longer present. Note, that the presence of even harmonics in the spectra is explained by a slight asymmetry in the input square wave.

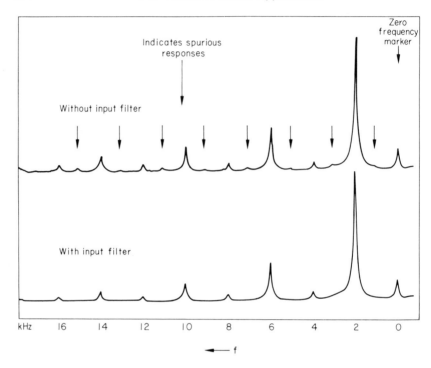

Fig. 6.27   Spectrum of 2kHz square wave

References

1. R.S. Brown and D. Sullivan *AD530 Complete Monolithic MDSSR* Technical Bulletin, Analog Devices (July 1971)

2. *Non-Linear Circuits Handbook*, Analog Devices (In Print)

3. D. Sheingold *Trigonometric Operations with the 433*, Analog Dialogue **Vol 6**, No 3

Exercises 6

6.1 A multiplier has a scaling factor $K = 1/10$. The input offset voltages associated with the X and Y inputs are:
$V_{iox} = 20\,\text{mV}$, $V_{ioy} = 15\,\text{mV}$.
What is the error due to the input offset voltages when input signal voltages $V_x = 6$ volts and $V_y = 8$ volts are applied to the multiplier? Neglect non-linearity and all sources of error other than those due to the input offset voltages.

6.2 The non-linearity specifications for a multiplier are given as, X input 1 per cent, Y input 0.4 per cent, the multiplier has a scaling factor $K = 1/10$. What is the maximum non-linearity error which can be expected for input signals 8 volts and 2 volts? What is the non-linearity error if input signals are connected in such a way as to minimise the error?

6.3 The 1594 multiplier is to be used to give a full scale output of magnitude 10 volts for maximum input signals $V_x = 1$ volt, $V_y = 0.5$ volts. What scaling factor is required? Suggest suitable values for resistors R_x, R_y and R_L (see figure 6.4). Assume $R_1 = 16\,\text{k}\Omega$ making $I_{cy} = 0.5$ milliamps.

6.4 Sketch the external connections to be made to a four quadrant multiplier (AD 532 type, see figure 6.9) in order to obtain an output signal of the form $V_o = -10\, V_z/V_x$ for positive values of V_x. Explain why small values of V_x lead to large errors and why the circuit will not operate with negative values of V_x;

6.5 A circuit of the type shown in figure 6.22 is to be used as an electrical analogue in order to generate the function
$$y = x - \frac{x^2}{2} + \frac{x^3}{3} - \frac{x^4}{4}$$
$$0 < x \leq 1$$
The multipliers to be used have a scaling factor $K = 1/10$ they are to be assumed to have a differential input capability. Choose values of the resistors R, R_a, R_b, R_c, R_d, for a full scale (± 10 volts) output for a 10 volt input signal. Sketch the circuit arrangement.

7. Phase Locked Loops

A phase locked loop, (PLL), is basically a closed loop feedback system, the action of which is to lock or synchronise the frequency of a controlled oscillator to that of an incoming signal. Phase lock principles are by no means new, synchronous reception of radio signals using PLL techniques was described as early as 1932. However, when implemented with discrete components, phase lock techniques involve circuits of considerable cost and complexity and for this reason their use in the past has been limited to specialised measurements requiring a high degree of noise immunity and very narrow bandwidth. The development of integrated circuit PLL's now makes it economically possible to employ the sophisticated phase lock techniques in a wide variety of signal detection and processing applications.

The reader who wishes to make an in-depth study of phase lock principles is referred to the several books on the subject [1,2,3] and to the extensive bibliography which they include. The treatment to be given here is a practically orientated approach intended to give a basic understanding of PLL's which is sufficient to enable the reader intelligently to use integrated circuit PLL's in a variety of applications.

7.1 Phase Locked Loop Building Blocks

In its simplest form a PLL consists of three functional blocks; a phase detector, a low pass filter and a voltage controlled oscillator, VCO. The basic loop may also contain an amplifier and the units are connected together in a closed loop as shown in figure 7.1. Integrated circuits for PLL's are available as separate functional blocks or as single devices in which all functions are included on a single chip.

Devices with performance characteristics fulfilling the requirements of those for the separate functional blocks have been described earlier in the book. Balanced modulators with circuitry of the type discussed in section 4.4 are used in many PLL's to perform the phase detector function in the loop. As an alternative some loops employ a digital type of phase detector, the Motorola MC 4344 device is an example of a digital phase detector designed for use in PLL's, the interested reader is referred to the manufacturers data sheet.

The VCO is usually the most critical block in a PLL. The frequency stability and the demodulation characteristics when the system is used for F.M. demodulation are normally determined by the VCO performance. Desirable characteristics for a VCO to be used in a PLL are

(1) Linearity of voltage to frequency conversion
(2) Frequency stability. Low frequency drift with temperature, supply voltage and time
(3) Ease of centre frequency setting with a minimum number of external components
(4) Wide range of frequency adjustment with control voltage
(5) A high frequency capability

The monolithic waveform generators which were discussed in section 3.3 can be used as VCO's, at frequencies up to about 1MHz, in separate block implementations of PLL applications. PLL's designed for use at higher frequencies generally employ a VCO which uses an emitter coupled multivibrator type of circuit. The Motorola MC 4324 device is an example of a voltage controlled multivibrator which can be used at frequencies up to about 30MHz.

7.2 The Phase Lock Loop Principle

Detailed analyses of the PLL as a feedback system are to be found in the literature and will not be repeated here. An understanding of the PLL principle can be gained from a qualitative description of the action of the loop which will now be given.

Referring to figure 7.1 the phase detector is assumed to exhibit a multiplier characteristic so that with no input signal applied to the system the output from the phase detector is zero. The error voltage applied as the control signal to the VCO is also zero and the VCO operates at its free running frequency, f_0' ($f_0' = \frac{\omega_0'}{2\pi}$), this frequency is referred to as the centre frequency. If an input signal is applied to the loop the phase detector produces an output signal which contains components at the sum and difference frequencies, $f_s + f_0$. If f_s is significantly different from f_0 both components are attenuated by the low pass filter, the frequency of the VCO is not changed and the loop does not acquire a lock.

A different situation exists when the frequency component $f_s \sim f_0$ lies within the pass band of the low pass filter, for this component of the phase detector output is then amplified and applied as a control signal to the VCO. This causes the VCO frequency to vary in a direction which reduces the

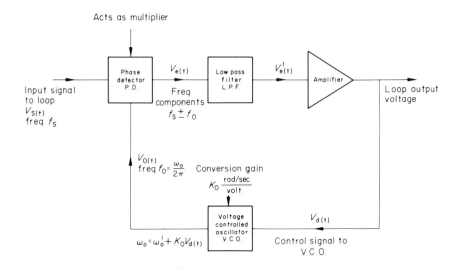

Fig. 7.1 Block diagram of basic PLL

frequency difference between f_s and f_0 and if f_s is sufficiently close to f_0 the feedback action of the loop causes the VCO to synchronise or 'lock' with the incoming signal. Once in lock the VCO frequency is identical to that of the input signal and the difference frequency component $f_s - f_0$ is a direct voltage of magnitude proportional to the cosine of the phase difference θ between the input signal and the VCO signal. The action of the loop is to cause θ to take on just that value which is required to generate the d.c. control voltage necessary to change the frequency of the VCO from its free running value to the frequency of the input signal. This action allows the PLL to 'track' any frequency changes of the input signal once lock has been acquired.

The range of frequencies over which a PLL can maintain lock with an input signal is called the 'lock range' of the system. As the input signal frequency is varied through the lock range the total variation possible for the phase angle θ is between $0°$ and $180°$ with a value $90°$ when the frequency is equal to the centre frequency. The VCO waveform used in most PLL's is a square wave, in which case the action of the phase detector is to multiply the input signal by a constant, (say A_d) and a unit amplitude square wave, (effectively multiplication by plus and minus A_d). Waveforms illustrating the phase detector output when the loop is in lock for various phase relationships between the input signal and VCO signal are shown in figure 7.2. The phase detector produces a

Linear Integrated Circuit Applications 211

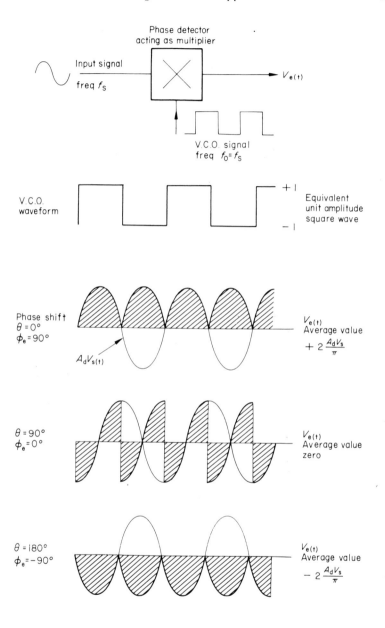

Fig. 7.2   Phase Detector output as a function of phase difference when loop is in lock

d.c. output component for input signals with frequencies equal to odd harmonics of the VCO frequency and the loop can thus lock to such signals. The phase detector d.c. output is lower for harmonic lock, and the lock range therefore decreases as higher order odd harmonics are used to achieve lock.

The range of frequencies over which a PLL can acquire lock is known as the 'capture range'. The greatest capture range possible in any loop is equal to the lock range for that loop but in general the capture range is less than the lock range. The capture of an input signal with a frequency in the capture range does not take place as soon as the signal is applied, a loop takes a finite time, called the 'pull-in' time, to establish lock. The capture process is complex and the pull-in time depends upon the initial frequency and phase difference between the input signal and the VCO signal as well as on the overall loop gain and the low pass filter bandwidth. In a specific loop with a fixed loop gain and fixed low pass filter bandwidth the pull-in time is likely to vary randomly with random initial phase relationships. The variation takes place, between some minimum pull-in time, corresponding to an initial phase relationship most favourable to capture and a maximum pull-in time, corresponding to the least favourable initial phase relationship.

A simple mathematical analysis of the capture process is not possible but some understanding of the mechanism may be gained from the following qualitative description. The time derivative of phase is equal to angular frequency and the frequency and phase errors in a loop may be related by

$$2\pi \, \Delta f = \Delta \omega = \frac{d\phi}{dt} \qquad (7.1)$$

Where $\Delta f = f_s \sim f_0$ and $\phi = \pi/2 + \theta$ represents the phase departure from a quadrature relationship. We remember that it is the difference frequency component in the output of the phase detector which, when passed by the low pass filter, acts as the control signal to the VCO and modulates its frequency. The result of this modulation is to make Δf vary with time during the capture process. If during the modulation f_0 moves towards f_s, Δf decreases with a corresponding decrease in $d\phi/dt$ and the output of the phase detector becomes a slowly varying function of time. If, on the other hand, f_0 moves away from f_s, Δf increases with a corresponding increase in $d\phi/dt$ and the output of the phase detector varies more rapidly with time. The overall effect is to make the difference frequency component waveform non sinusoidal, its asymmetrical nature gives it a d.c. component which forces the average value of the VCO frequency towards that of the input signal and finally results in the acquisition of lock. Under the conditions discussed the phase detector may give a beat note waveform during capture of the type sketched in figure 7.3. The exact capture

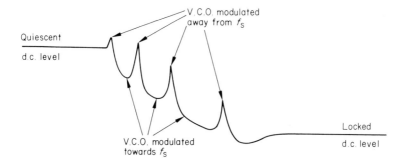

Fig. 7.3　　Phase detector beat note during capture process

transient, in a practical situation, varies with initial phase relationships and loop circuit conditions. Note that it is possible under certain loop conditions for the pull-in time to be less than the period of the beat note and lock is then acquired rapidly without an oscillatory phase error transient. Some experimentally observed capture transients are shown in figure 7.9 of the next section.

Before continuing with a further treatment of PLL characteristics it is appropriate to introduce a practical test circuit. The reader is urged to perform this practical evaluation of the PLL principle for himself; it should help to clarify the discussion and the terminology that has been introduced. The terms that have been introduced in this section are restated for additional emphasis.

Free-Running Frequency, Centre Frequency, $(f_0{'}, \omega_0{'} = 2\pi f_0{'})$. This is the frequency at which the loop VCO operates when not locked to an input signal.

Lock Range, Tracking Range, Hold-in Range, (f_L). The range of frequencies in the vicinity of $f_0{'}$ over which the PLL, once locked to the input signal, will remain in lock. f_L represents the value of $F_s - f_0{'}$ at which the loop fails to track f_s. Note that the complete lock range is centred about $f_0{'}$ and some sources define $2f_L$ as the lock range. We shall treat lock range, tracking range and hold-in range as the same quantities equal to f_L.

Capture Range, Pull-in Range, Acquisition Range, (f_c). The maximum initial frequency difference between the input signal and the VCO for which the loop can acquire lock. Capture can take place for a range of frequencies f_c either side of $f_0{'}$, a total capture range of $2f_c$ centred about $f_0{'}$.

Pull-in Time, Lock-up Time

The time taken for a PLL to acquire lock when an input signal with frequency within its capture range is suddenly applied to the loop.

7.3 Measurement of Lock and Capture Range – Display of Capture Transient

An experimental evaluation of PLL action can be made in a variety of different ways. A loop can be connected up using i.c. devices as the separate functional blocks for the loop, or, more conveniently, a single chip PLL can be investigated. A single chip PLL was used by the author for the purpose of the test measurements to be described. It was a Signetics NE 565 device which is a general purpose PLL suitable for use at frequencies up to 1 MHz.

The measurement of the lock and capture range for a PLL requires the use of a variable frequency signal source as an input signal for the loop and a measurement of the loop error voltage for various values of the input signal frequency. If a voltage swept signal source is not available the measurements

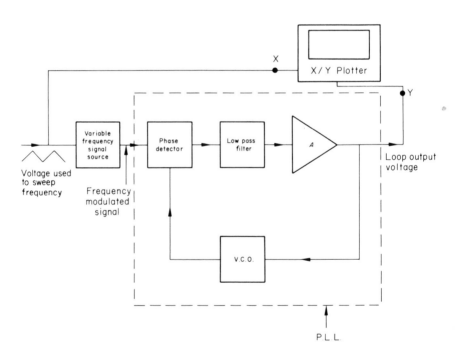

Fig. 7.4   Measurement of loop transfer characteristic

can be made point by point with the input signal frequency varied by manual adjustment of the control dial on the signal source. However, if a series of measurements are to be made, in order to investigate the factors which influence lock and capture range, it is desirable to arrange a test circuit which gives an X/Y display of the input frequency to error voltage transfer characteristic of the loop. The principle of the test is illustrated by the block diagram in figure 7.4 and a practical test circuit is shown in figure 7.5. An X/Y recorder is used to produce the transfer curves but, if required, an oscilloscope can be used instead with the appropriate deflecting signals applied to its horizontal and vertical deflection channels.

In the NE 565 device the free running frequency of the VCO is determined by the values of an external resistor R_1 connected between pin 8 and the positive supply line and an external capacitor C_1 connected between pin 9 and the negative supply line. A capacitor of value typically 0.001 μF is normally connected between pins 7 and 8 to eliminate possible oscillation in the VCO voltage controlled current source. The manufacturers give the approximate design equation $f_0' \cong \dfrac{1.2}{4 R_1 C_1}$, it is usual to fix C_1 and experimentally trim the value of R_1 for a particular required value of f_0'. Values of R_1 in the range $2 - 20\text{k}\Omega$ are recommended with an optimum value of order $4\text{k}\Omega$.

The square wave output signal of the VCO is available at pin 4 and in order to close the loop pin 4 must be connected externally to the phase detector input at pin 5. The amplified loop error voltage, which is applied as the control signal to the VCO, is available at pin 7, this signal is referenced to the positive supply line. A reference voltage which is nominally equal to the voltage at pin 7 is available at pin 6 and this allows differential stages to be both biased and driven by connecting them to pins 6 and 7. The loop low pass filter is formed by an internal resistor of value typically 3.6 kΩ and an external capacitor (or capacitor and resistor in series), connected between pin 7 and the positive supply line. The signal inputs to the phase detector are differential (pins 2 and 3), and the d.c. level at these two pins must be made the same. If dual power supplies are used it is simplest to bias pins 2 and 3 at the potential of the common power supply line. With single supply operation they should be biased to a level in the lower half of the total supply voltage by means of a suitable potential divider.

A circuit schematic for the NE 565 device is given in figure 7.6. The reader should recognise the circuitry of the VCO section as that of the monolithic waveform generator device, type NE 566 discussed in section 3.3. The phase detector section uses balanced modulator circuitry of the type discussed in section 4.4.

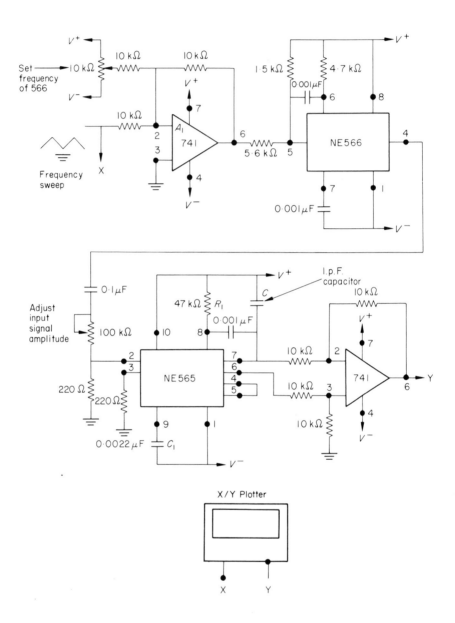

Fig. 7.5   Practical test circuit for investigation of loop transfer characteristic

Linear Integrated Circuit Applications 217

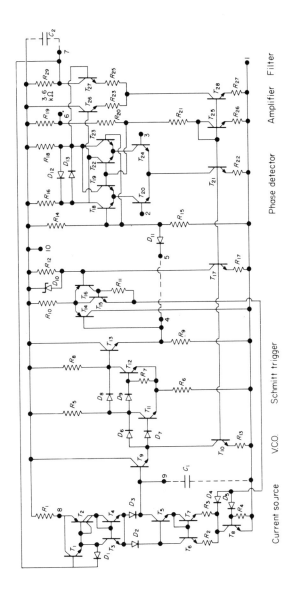

Fig. 7.6   NE565 circuit schematic

218 Linear Integrated Circuit Applications

In the test circuit of figure 7.5 the input signal to the loop is provided by a type NE 566 device, thus allowing for a frequency sweep by means of an externally applied sweep voltage. The amplifier A_1 allows an earth-referred sweep voltage to be used. This may be derived from a potentiometer, for a manual sweep, or alternatively a low frequency triangular wave derived from a waveform generator can be used as the frequency sweep signal. The sweep frequency must be very much less than the centre frequency because of lock-up time duration and variation. If sweep frequency is made too high, capture range will appear to be a function of sweep frequency.

Typical transfer curves obtained in the test are shown in figure 7.7 and 7.8 for centre frequencies of 22.4 kHz and 47.2 kHz respectively. The three curves given in each figure were obtained with values $1\,\mu F$, $0.1\,\mu F$ and $0.01\,\mu F$

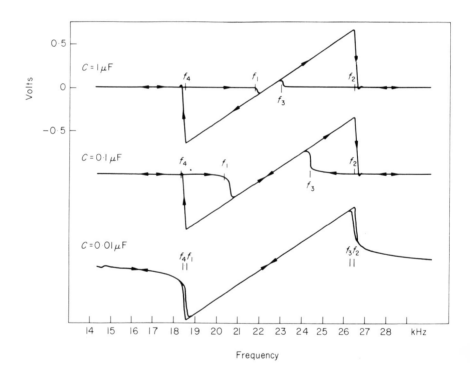

Fig. 7.7   Loop transfer curves; centre frequency 22.4 kHz

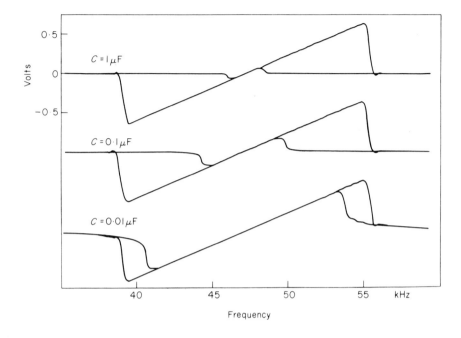

Fig. 7.8   Loop transfer curves; centre frequency 47.2 kHz

used for the low pass filter capacitor. A digital frequency meter was used to calibrate the horizontal axis of the display in terms of frequency prior to obtaining the transfer curves. In each curve the frequency is swept from a frequency well below the centre frequency to one considerably higher than the centre frequency and back again. Starting at the low frequency end the loop does not respond to the input signal until its frequency reaches a value f_1 corresponding to the lower edge of the capture range. The loop then locks to the input causing a sudden change in the loop error voltage. The loop tracks the input signal until its frequency reaches the value f_2 corresponding to the upper edge of the lock range. The loop then loses lock and the error voltage returns to zero. On the return frequency sweep the loop acquires lock again at the frequency f_3 corresponding to the upper edge of the capture range and tracks down to the frequency f_4, corresponding to the lower edge of the lock range. Measurements of the frequencies f_1, f_2, f_3 and f_4 give a direct indication of the lock and capture range of each loop from the relationships

$$2f_L = f_2 - f_4 \: : \: 2f_c = f_3 - f_1$$

The linearity of the frequency-to-voltage transfer characteristic of a PLL is entirely dependent upon the linearity of the voltage-to-frequency conversion performed by the VCO. If we let K_0 represent the VCO conversion gain in radians per second/volt the slope of the transfer characteristic is equal to $2\pi/K_0$ volts per cycle per sec.

Many other useful measurements can be made with the test circuit of figure 7.5. It is suggested that the loop transfer characteristic for harmonic lock be measured and compared with that for fundamental lock, the dependence of lock and capture range on input signal amplitude should also be investigated. The curves in figures 7.7 and 7.8 were obtained for the phase detector used in the limiting mode, ($V_s \gg 200$ mV p-p), when the lock and capture range are independent of signal amplitude (see later).

The nature of the loop capture transient can be investigated with the test circuit. This is done by applying a control voltage step to the NE 566 signal source, instead of the slow sweep, so as suddenly to switch its frequency from one outside the lock and capture range of the loop to a frequency within the capture range. The capture transient, as exhibited by the loop error voltage, is displayed by an oscilloscope. Three capture transients which were obtained in this way are shown in figure 7.9. Single shot operation of the oscilloscope time base was used in order to display the transients.

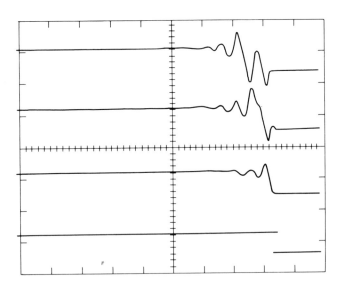

Fig. 7.9   Error voltage capture transients; time scale 1 ms/division

Linear Integrated Circuit Applications   221

Having achieved a basic practical familiarity with PLL action, as a result of performing the test measurements which have been suggested, the reader should more readily appreciate the further analysis of loop action which is now to be presented. The test circuit should be returned to whenever an experimental verification of some aspect of loop performance is required.

7.4 Parameters determining Lock and Capture Range

In section 7.2 we discussed the way in which a PLL, when in lock, generates a phase difference (phase error), between the input and VCO signals and how this phase error, in the form of a proportional voltage, controls the VCO frequency in such a way as to maintain synchronism with the input signal. In order to analyse the process further it is first necessary to examine in greater detail the parameters which are used to characterise the performance of the individual functional blocks in the loop.

7.4.1 *Phase Detector Conversion Gain* K_d

The phase detector in a phase locked loop performs a multiplier operation. The performance of a linear multiplier is characterised by its so called scaling factor (see section 6.1) but in a PLL operation the amplitudes of the input signals applied to the phase detector are generally greater than those necessary for a linear scaling operation and input signal limiting occurs. Under these conditions the phase detector output contains frequency components at the sum and difference frequencies of harmonic frequencies of the input and VCO frequencies. When the loop is in lock these frequency components do not affect the basic action of the loop and they are rejected by the low pass filter in the loop. However, input signal amplitude limiting makes the magnitude of the 'd.c.' component of the phase detector output, which is present under locked conditions, independent of input signal amplitudes and dependent only upon the cosine of the phase difference between the input signals.

It is usual to express phase difference in terms of the departure from a quadrature relationship between the input and VCO signals and when so expressed the phase detector 'd.c.' output is determined by the relationship

$$V_e = K_d \sin \phi_e \qquad (7.2)$$

or $V_e \cong K_d \phi_e$ for small values of ϕ_e.

K_d is called the conversion gain of the phase detector in volts/radian. Note that K_d is only a constant for values of the input signal amplitude greater than some minimum value, (> 200mV p-p for the NE 565 device). K_d decreases when the input signal amplitude is decreased below this value.

7.4.2 Low Pass Filter and Amplifier

The function of the low pass filter is to attenuate the high frequency components in the phase detector output signal. The low frequency components are passed by the filter, amplified by the loop amplifier and applied as a control to the VCO. The filter affects the capture characteristics of the loop and the dynamic response of the loop to changes in the input signal frequency. When the loop is in lock an input signal of fixed frequency gives a constant phase error signal which is not attenuated by the low pass filter. If the phase of the input signal is modulated a corresponding modulation in the phase error is produced and this is attenuated by the low pass filter to an extent determined by the modulating frequency.

Note: frequency is equal to the time derivative of phase, frequency modulation of the input signal is thus accompanied by a phase modulation of the input signal.

In the 'language' of electrical network theory the performance of a filter may be characterised by its transfer operator, $F(p)$, (See Appendix A). Expressed in these terms the output signal from the loop low pass filter may be written as

$$V'_e(t) = F(p) V_e(t) \tag{7.3}$$

This signal is amplified by the loop amplifier before being applied as the control signal to the VCO. The loop amplifier has a d.c. response and its upper bandwidth limit extends to frequencies higher than those present in $V'_e(t)$. Its performance is thus adequately represented by its gain A (a real number) and we may write

$$V_d(t) = A\, V'_e(t) \tag{7.4}$$

7.4.3 VCO Conversion Gain K_0

In the absence of a control signal the VCO free runs, producing a waveform with a fundamental angular frequency ω_0'[1]. A non-zero control signal causes the VCO frequency to change from its free running value and time variations of the control signal produce corresponding time variations in the VCO frequency. Assuming a linear performance the relationship between VCO frequency and control voltage may be expressed as

$$\omega_{0(t)} = \omega_0' + K_0 V_{d(t)} \tag{7.5}$$

$\omega_{0(t)}$ is the instantaneous angular frequency of the VCO output for an instantaneous control voltage $V_{d(t)}$.

The constant K_0 is called the VCO conversion gain and is expressed in rad/sec/volt. In a practical VCO the value of K_0 is dependent upon the value used for ω_0'.

The fundamental component of the VCO output waveform may be written as

$$V_{o(t)} = V_o \sin(\omega_o' t + \phi_{o(t)}) \quad (7.6)$$

Remembering that frequency is the time derivative of phase we may write

$$\frac{d}{dt}(\omega_o' t + \phi_{o(t)}) = \omega_{o(t)} = \omega_o' + K_o V_{d(t)}$$

Thus

$$\frac{d\phi_{o(t)}}{dt} = K_o V_{d(t)} \quad (7.7)$$

and

$$\phi_{o(t)} = K_o \int_0^t V_{d(t)} \, dt$$

or

$$\phi_{o(t)} = \frac{K_o}{p} V_{d(t)} \quad (7.8)$$

If we regard phase as the loop variable the VCO acts as an integrator in the loop.

Having defined the performance parameters of the functional elements in the loop we are now in a position to examine the way in which the loop characteristics are related to the values of these parameters. From the discussion in section 7.2 and the diagrammatic treatment of figure 7.2, it is evident that the maximum phase error (the phase departure from a quadiature relationship between input signal and VCO signal), that can exist in the loop for the loop to remain in lock is $\phi_e = \pm 90°$.

The control voltage is

$$V_{d(t)} = A \, F_{(p)} K_d \sin \phi_{e(t)}$$

Assuming 'd.c.' tracking ϕ_e does not vary with time, $F_{(p)} = 1$ and the maximum value of V_d is

$$|V_d|_{max} = A K_d$$

Using equation 7.5 the maximum tracking range ω_L is

$$\omega_L = K_o |V_d|_{max}$$

or $\omega_L = K_o A K_d = K_V \quad (7.9)$

where K_V is the total gain round the loop. The tracking range is numerically equal to the d.c. gain round the loop and does not depend upon the characteristics of the low pass filter.

Loop capture is a transient condition which is difficult to analyse, we make no attempt to perform this analysis but quote an approximate expression for capture range.[4]

$$\omega_c = K_0 K_d A |F_{(j\omega_c)}| \qquad (7.10)$$

$|F_{(j\omega_c)}|$ is the magnitude of the low pass filter response to a sinusoidal signal at an angular frequency ω_c.

Note that at all times $|F_{(j\omega)}| \leq 1$ so that capture range is always less than lock range. If the low pass filter is omitted lock and capture range are equal.

7.5 Dynamic Behaviour of the Locked Loop

The factors influencing the dynamic behaviour of a PLL when in lock are now examined. For the purpose of this analysis consider an input signal of the form

$$V_{s(t)} = V_s \cos(\omega_0' t + \phi_{s(t)}) \qquad (7.11)$$

V_s is the amplitude of the input signal which is assumed to be constant. $\phi_{s(t)}$ is the input signal phase modulation.

The loop is assumed to be in lock and the fundamental component of the VCO output may be written as (equation 7.6)

$$V_{0(t)} = V_0 \sin(\omega_0' t + \phi_{0(t)})$$

$\phi_{0(t)}$ is the phase modulation of the VCO signal.

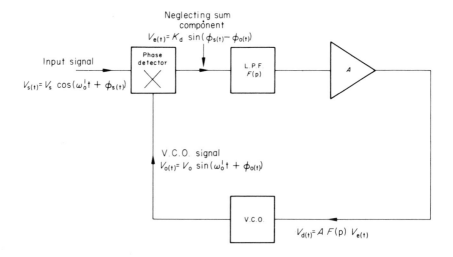

Fig. 7.10   Signals present in locked loop

Note: The form of the equations is chosen for the quadrature relationship between input signal and VCO signal which exists when the input signal frequency is equal to the VCO centre frequency. The signals present at various points in the loop are illustrated in figure 7.10.

We continue by considering an input signal in which $\phi_{s(t)}$ is of the form

$$\phi_{s(t)} = \Delta \omega_s t + \phi'_{s(t)} \qquad (7.12)$$

$\Delta \omega_s$ represents a static input frequency offset from the loop centre frequency and $\phi^1_{s(t)}$ represents a dynamic signal modulation. Assuming lock is maintained, the loop responds to the input signal phase modulation by producing a phase modulation of the VCO signal and

$$\phi_{o(t)} = \Delta \omega_s t + \phi_k + \phi'_{o(t)} \qquad (7.13)$$

ϕ_k is a constant phase error generated, by the loop because of $\Delta \omega_s t$ and $\phi'_{o(t)}$ is a dynamic response of the loop to the signal modulation.

Neglecting the signal component with frequency equal to the sum of the input and VCO frequencies the phase detector output may be written as, (equation 7.2)

$$V_{e(t)} = K_d \sin \phi_{e(t)}$$

$$= K_d \sin (\phi_{s(t)} - \phi_{o(t)})$$

Substitution for $\phi_{s(t)}$ and $\phi_{o(t)}$ from equation 7.12 and 7.13

$$V_{e(t)} = K_d \sin [(\phi'_{s(t)} - \phi'_{ot}) - \phi_k]$$

If the dynamic phase error $\phi'_{e(t)} = \phi'_{s(t)} - \phi'_{o(t)}$ is small we may write

$$\sin (\phi'_{e(t)} - \phi_K) = \phi'_{e(t)} \cos \phi_K - \sin \phi_K$$

and substitution in equations 7.3 and 7.4 gives

$$V_{d(t)} = F_{(p)} A K_d [\phi'_{e(t)} \cos \phi_K - \sin \phi_K]$$

Differentiating equation 7.13 gives

$$\frac{d \phi_{o(t)}}{dt} = \Delta \omega_s + \frac{d \phi'_{o(t)}}{dt}$$

and substitution in equation 7.7 yields

$$\Delta \omega_s + \frac{d \phi'_{o(t)}}{dt} = F_{(p)} A K_d K_0 [\phi'_{e(t)} \cos \phi_K - \sin \phi_K] \qquad (7.14)$$

Equation 7.13 may be written as two parts

$$\Delta \omega_s = - F_{(p)} A K_d K_0 \sin \phi_K \qquad (7.15)$$

$$\frac{d \phi'_{o(t)}}{dt} = F_{(p)} A K_d K_0 \phi'_{e(t)} \cos \phi_K \qquad (7.16)$$

Equation 7.15 represents the static (or d.c.) tracking equation. Equation 7.16 is a linear approximation representing the dynamic tracking performance of the loop. It is valid for small values of the dynamic phase error $\phi'_{e(t)}$. For purely static frequency offsets $F_{(p)} = 1$ and equation 7.15 gives

$$\omega_L = |\Delta \omega_s|_{max} = A\, K_d\, K_0\, \text{Sin}\, 90°$$

which is in agreement with the result previously derived in equation 7.9

Equation 7.16 describes the behaviour of a linear closed loop control system in which phase angle is the variable. Not all readers will be familiar with the theory of linear control systems but most are likely to have a familiarity with feedback theory as applied to amplifiers (to op. amps.). We therefore interpret equation 7.16 in terms of a feedback amplifier in which phase is the input variable, this interpretation is illustrated diagramatically in figure 7.11.

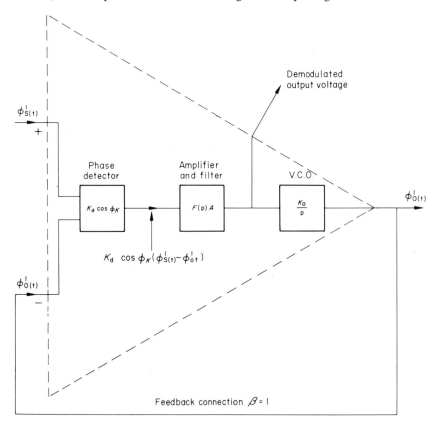

Fig. 7.11 Locked loop; phase amplifier representation; a phase follower

The operator $p = \dfrac{d}{dt}$ and equation 7.16 may be written as

$$p \phi'_{o(t)} = F_{(p)} \, A \, K_d \, K_0 \, \text{Cos} \, \phi_K \, \phi'_{e(t)}$$

or

$$\phi'_{o(t)} = [K_d \, \text{Cos} \, \phi_K] \, F_{(p)} \, A \, \frac{K_0}{p} \, \phi'_{e(t)} \qquad (7.17)$$

Output action of	action	loop	action of	input (phase)
(phase) phase	of	ampl.	VCO	error
signal detector	filter		(integ.)	signal

If we imagine the loop in our 'phase' amplifier to be broken we may define its open loop gain as

$$A_{OL} = \frac{\phi'_{o(t)}}{\phi'_{s(t)OL}} = K_d \, \text{Cos} \, \phi_K \, F_{(p)} \, A \, \frac{K_0}{p} = \frac{F_{(p)} K_v}{p} \, \text{Cos} \, \phi_K \qquad (7.18)$$

Where we write $\phi'_{s(t)} = \phi'_{e(t)}$ for the open loop system.

When the loop is closed the feedback fraction $\beta = 1$ and by the usual negative feedback theory we may write

$$A_{CL} = \frac{\phi'_{o(t)}}{\phi'_{s(t)CL}} = \frac{A_{OL}}{1 + \beta A_{OL}}$$

or

$$\frac{\phi'_{o(t)}}{\phi'_{s(t)CL}} = \frac{K_v \, F_{(p)} \, \text{Cos} \, \phi_K}{p + K_v \, F_{(p)} \, \text{Cos} \, \phi_K} \qquad (7.19)$$

It is evident from equation 7.19 that the dynamic response of a locked loop is strongly dependent upon the nature of the loop low pass filter. The dynamic behaviour to be expected with different types of filter will now be discussed.

7.5.1 First Order Loop

The simplest type of loop, the so called first order loop, is one in which no low pass filter is used. Making $F_{(p)} = 1$; the expression for the closed loop 'phase' gain (equation 7.19) becomes

$$A_{CL(p)} = \frac{K_v \, \text{Cos} \, \phi_K}{p + K_v \, \text{Cos} \, \phi_K} \qquad (7.20)$$

Equation 7.19 represents a first order low pass transfer function (see appendix equation A4). Writing $p = j\omega$, the response of the loop to a steady state sinusoidal modulation of the input phase is given as

$$A_{CL(j\omega)} = \frac{\phi'_o(j\omega)}{\phi'_s(j\omega)_{CL}} = \frac{1}{1 + j \dfrac{\omega_m}{K_v \, \text{Cos} \, \phi_K}} \qquad (7.21)$$

The upper bandwidth limit occurs at a modulation frequency $\omega_{m\,3dB} = K_v \cos\phi_K$. Bode plots are used in discussing the behaviour of feedback amplifiers, the bode plots for our first order 'phase' amplifier are shown in figure 7.12.

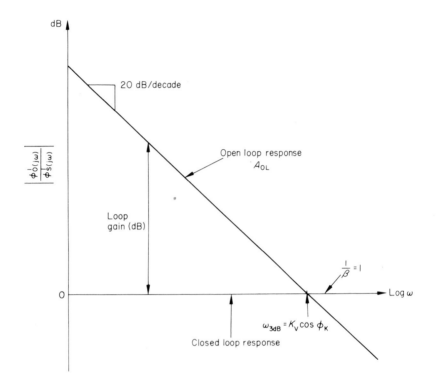

Fig. 7.12   Bode plot for phase amplifier; first order loop

Note that at the intersection of $1/\beta$ and the open loop response, the rate of closure is 20 dB per decade giving a stable closed loop system with no response peaking.

7.5.2 Second Order Loop

We now consider the addition of a simple first order low pass filter to the loop, this gives the loop itself a second order characteristic. In the test circuit used to investigate the NE 565 device an external capacitor connected to pin 7 of the device forms a first order low pass filter with a resistor inside the device.

The transfer characteristic for a first order low pass filter may be written as

$$F_{(p)} = \frac{1}{1 + \frac{p}{\omega_1}}$$

Where $\omega_1 = \frac{1}{CR_1} = \frac{1}{r_1}$

R_1 is the value of the internal resistor, C is the value of the external capacitor. With this type of transfer characteristic equation 7.18 now becomes

$$A_{CL} = \frac{\phi_o'(t)}{\phi_s'(t)} = \frac{K_v \cos \phi_K}{1 + \frac{p}{\omega_1}\left[p + \frac{K_v \cos \phi_K}{1 + \frac{p}{\omega_1}}\right]}$$

which may be written as

$$A_{CL} = \frac{1}{1 + \frac{p}{K_v \cos \phi_K} + \frac{p^2}{\omega_1 K_v \cos \phi_K}} \qquad (7.22)$$

Equation 7.22 should be compared with the general expression for a second order low pass transfer function which is given in the Appendix (equation A7). Equation 7.22 represents a second order system with damping factor

$$\zeta = \tfrac{1}{2}\left(\frac{\omega_1}{K_v \cos \phi_K}\right)^{1/2} \qquad (7.23)$$

and natural frequency

$$\omega_0 = \left(\omega_1 K_v \cos \phi_K\right)^{1/2} \qquad (7.24)$$

Bode plots for the system are shown in figure 7.13. Note that at the intersection of $1/\beta$ and the open loop response the rate of closure is greater than 20 dB per decade corresponding to a phase shift round the loop of greater than 90°. The closed loop gain may be expected to show a definite peaking. Corresponding to the peaking in steady state response we may expect an overshoot and ringing in the transient response of the loop in response to a step change in the phase of the input signal. The extent of the steady state gain peaking and transient ringing is determined by the value of the damping factor. Equations governing the behaviour are included in the general

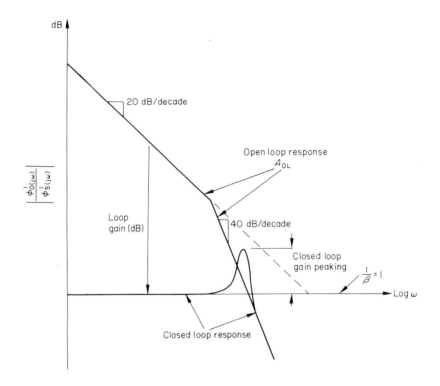

Fig. 7.13   Bode plot; second order loop

treatment of a second order system which is given in the Appendix. Note that for a fixed value of K_v the loop becomes more and more underdamped as the filter time constant $\tau_1 = 1/\omega_1$ is increased. For a fixed filter time constant damping is reduced by an increase in the gain K_v (see equation 7.23).

A very lightly damped loop is not desirable and in practice may cause the loop to break into sustained oscillations. The instability problem, if it exists, can be overcome by using a lag lead type of filter. A series capacitor/resistor combination connected to pin 7 of the NE 565 device may be used to form a lag lead filter with the internal resistor in the device. When used with this type of filter the demodulated output signal of the loop is taken from across the capacitor C instead of from pin 7. The transfer characteristic of a lag lead

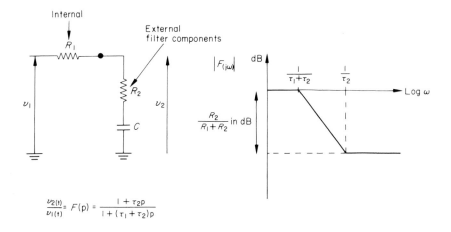

Fig. 7.14   Lag-lead filter

low pass filter may be written as

$$F_{(p)} = \frac{1 + \tau_1 p}{1 + (\tau_1 + \tau_2) p}$$

where $\tau_1 = CR_1$ $\tau_2 = CR_2$ (See figure 7.14)

The bode plots for the phase amplifier when used with this type of filter take on the form shown in figure 7.15. Proper choice of R_2 fixes the second break in the open loop response at the unity gain crossover frequency

$$\frac{K_v Cos\phi_K + \dfrac{1}{C(R_1 + R_2)}}{2}$$

giving a phase margin of 45° with 3 dB of peaking in the steady state response. The use of a larger value for R_2 gives a greater phase margin and as R_2 is increased the behaviour of the loop approaches that of the first order loop. If R_2 is decreased below its proper value phase margin is decreased and the system becomes lightly damped with a corresponding increase in gain peaking. Expressions for the natural frequency and damping factor of the loop when

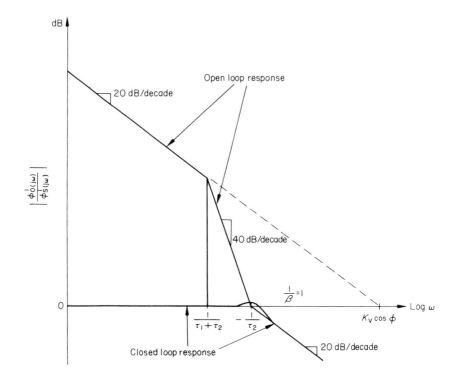

Fig. 7.15   Bode plot with lag-lead filter

used with a lag lead type of filter are

$$\omega_0 = \left(\frac{K_v \cos \phi_K}{\tau_1 + \tau_2}\right)^{1/2} \tag{7.25}$$

$$\zeta = \tfrac{1}{2}\omega_0 \left[\tau_2 + \frac{1}{K_v \cos \phi K}\right] \tag{7.26}$$

In summary we see that choice of filter type and filter component magnitudes offers the designer a powerful but simple (only a single capacitor or a capacitor and resistor are required) technique for controlling the dynamic performance characteristics of a phase locked loop. An arbitrarily narrow

bandwidth can be fixed by appropriate choice of τ_1 yet the loop can still be
adequately damped by using a lag lead type of filter. The use of a narrow
bandwidth, (long time constant filter) improves the interference and noise
rejection properties of the loop but at the same time the effect of a long time
constant filter in decreasing capture range and increasing lock up time should
not be forgotten. In F.M. demodulation applications the bandwidth must
clearly be designed so that it is wide enough to allow a faithful
reproduction of the highest modulation frequencies present.

7.5.3 Measurement of Dynamic Response

A practical investigation of the dynamic behaviour of a phase locked loop can
be readily carried out using the test circuit previously introduced in figure 7.4.
The test arrangement is illustrated again in block schematic form in figure 7.16.
In order to measure the steady state response of the loop a sinusoidal
modulating signal is applied to the voltage controlled generator and the
demodulated output signal produced by the loop is measured for different
values of the modulating frequency. In the case of a first order loop, where no
low pass filter is used, both sum and difference frequency components are

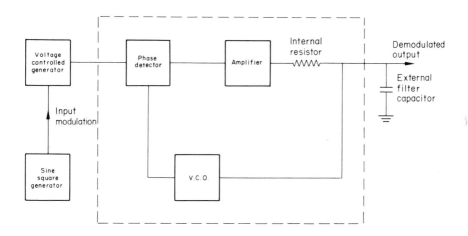

Fig. 7.16   Test circuit arrangement for dynamic response measurements

present in the phase detector output. The sum component must be filtered outside the loop if the difference frequency component (demodulated F.M.) is to be measured.

An expression for the frequency modulated input signal which is applied to the loop maybe written as

$$V_{s(t)} = V_s \cos[(w_o^1 + \Delta\omega_s)t + \frac{\Delta\omega}{\omega_m}\sin\omega_m t] \qquad (7.27)$$

$\Delta\omega_s$ represents a static frequency offset from the loop centre frequency, $\Delta\omega \cos\omega_m t$ represents the dynamic frequency modulation. Comparison with equations 7.10 and 7.11 allows us to write an expression for the dynamic signal phase modulation as

$$\phi'_{s(t)} = \frac{\Delta\omega}{\omega_m}\sin\omega_m t \qquad (7.28)$$

The dynamic variation of the VCO control voltage in response to the input signal frequency modulation may be written as

$$V_d'(t) = \frac{p}{K_0}\phi_0'(t)$$

and by the use of equations 7.19 and 7.28 we obtain

$$V_d'(t) = \frac{p}{K_0} \frac{K_v F_{(p)} \cos\phi_K}{p + K_v F_{(p)} \cos\phi_K}\phi'_{s(t)}$$

or

$$V_d'(t) = A_{CL} \frac{\Delta\omega \cos\omega_m t}{K_0}$$

It is this signal at the modulation frequency which represents the demodulated output of the loop.

Measurements of the transient response of the loop to sudden changes in the phase of the input signal may be made by applying a small amplitude squarewave to the voltage control generator as a modulating signal. An oscilloscope is then used to monitor the demodulated output given by the loop, waveforms obtained with the test circuit of figure 7.16 for two different values of filter capacitor (different damping) are shown in figures 7.17 and 7.18.

Measurements of the steady state gain peaking and transient response overshoot and ringing may both be used to estimate values for the loop damping and the natural frequency of the loop, (see Appendix A for details of second order system equations).

Linear Integrated Circuit Applications

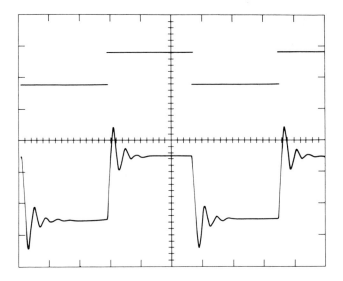

Fig. 7.17 Transient response of loop with filter capacitor 0.047 μF upper trace, input modulating signal 0.5 V/division; lower trace, demodulated output 0.2 V/division; time scale 1 ms/division

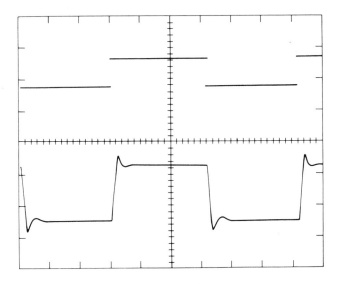

Fig. 7.18 Transient response with filter capacitor = 0.01 μF time scale = 0.5 ms/division

7.6 Modifying the Loop Characteristics

Some of the ways in which the characteristics of a loop can be varied have already been discussed. Thus in section 7.3 we saw how the value of the low pass filter capacitor determines the loop capture range and in section 7.5 the way in which the low pass filter controls the dynamic behaviour of the loop was discussed. Further control of loop characteristics can be achieved by modifying the loop so as to change the effective value of K_v the total gain round the loop. It should be remembered that: $K_v = K_o K_d A$ and in an i.c. phase locked loop, in which the feedback from the phase detector to the VCO is internal, it is not immediately obvious how the value of K_v can be changed.

The solution to the problem of changing the effective value of K_v in the case of i.c. phase locked loops can usually be found in the addition of external circuitry which modifies the effective value of K_0 the VCO conversion gain. We consider such modifications in relationship to the i.c. phase locked loop that we have used for the purpose of experimental evaluations.

In the 565 device the VCO conversion gain is determined by the product $C_1 R_1$ of the values of the external timing capacitor and the external resistor R_1. The circuitry of the VCO in the 565 is in fact that of the monolithic waveform generator device type N E 566 which was discussed in section 3.2.1. The key to any modification of the VCO is the realisation that it is the current into pin 8, normally supplied through R_1, which charges the timing capacitor and controls the VCO frequency. The addition of a constant current into pin 8, over and above that supplied through R_1, has the effect of reducing the percentage effect of the internal feedback voltage in changing the frequency of the VCO. The value of K_0 (expressed in units of f_0) is reduced by the addition of a constant current into pin 8.

7.6.1. *Reducing the Lock Range of the 565 (reference 5)*

A circuit for injecting a constant current into pin 8 of the 565 device and thereby reducing its lock range is illustrated in figure 7.19. The diode-connected transistor T_2 in the circuit is used to compensate for the temperature dependence of the transistor T_1. The constant current which is injected into pin 8 is determined approximately by the equation

$$I_c = \frac{V_s}{R_B}$$

Under normal operating conditions the VCO frequency is determined

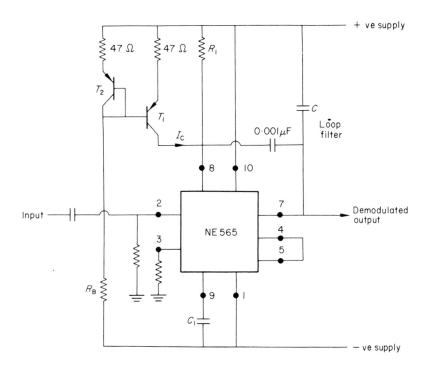

Fig. 7.19   Reduced lock range

by the relationship

$$f_0 \simeq \frac{5}{2} \frac{V_s - V_c}{V_s C_1 R_1} \qquad (7.29)$$
(see section 3.3.1)

In the absence of an input signal to the loop the 565 circuitry establishes a control voltage V_c which is directly proportional to the total supply voltage V_s. This makes the loop centre frequency independent of the supply voltage, the centre frequency design equation is given as

$$f_0^1 = \frac{1.2}{4 R_1 C_1}$$

Differentiation of equation 7.29 allows us to find an expression for the VCO conversion gain. Thus

$$K_0 = 2\pi \frac{\partial f}{\partial V_c} = \frac{5}{2} \frac{1}{V_s C_1 R_1} 2\pi \cong \frac{50 f_0'}{V_s} \text{ rad/sec/volt}$$

With the addition of a constant current into pin 8 the loop centre frequency becomes

$$f_0'' = \frac{1.2}{4 R_1 C_1} + \frac{5}{2} \frac{I_c}{V_s C_1}$$

$$= \frac{1.2 + 10 \frac{R_1}{R_B}}{4 R_1 C_1}$$

Expressed in terms of this new centre frequency the VCO conversion gain is now

$$K_0 \cong \frac{63 \; f_0''}{V_s \left[1.2 + 10 \frac{R_1}{R_B} \right]}$$

For a particular centre frequency the new look range is thus

$$\frac{63}{50 \left[1.2 + 10 \frac{R_1}{R_B} \right]} \times 100\%$$

of the lock range of the unmodified circuit.

7.6.2. *Increasing the Lock range of the 565*

The lock range of a phase locked loop can be extended by increasing K_v the gain round the loop. In an i.c. phase locked loop in which the feedback is internal it is not possible physically to break the loop and introduce extra gain. However, a way can usually be found of introducing extra external gain in such a way as to override the internal feedback connection. In the 565 device this can be done by supplying all the VCO charging current, into pin 8, by way of an external current source which is externally controlled by the loop error voltage. A circuit arrangement for this purpose is shown in figure 7.20. In this circuit the VCO frequency is determined by the current I_R and the value of the timing capacitor C_1; the internal feedback path is

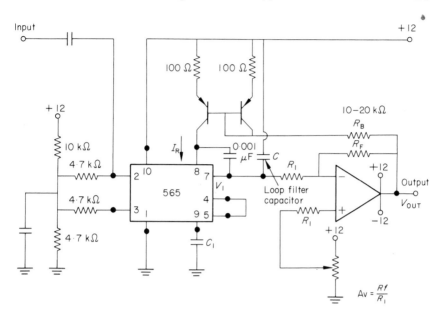

Fig. 7.20 Increased loop gain and lock range for 565

inoperative. The equation for the VCO frequency is

$$f = \frac{5}{2} \frac{I_R}{V_s C_1} \quad \text{where } I_R = \frac{V_0}{R_B}$$

Note that the VCO frequency is no longer independent of the supply voltage. Adjustment of the voltage to the non-inverting input terminal of the operational amplifier allows a centre frequency control. Loop gain and lock range can be controlled by the operational amplifier gain setting resistor R_f. The scheme allows the addition of sufficient gain to the loop for the loop to stay in lock over 100 to 1 frequency range. For small frequency variations of the input signal frequency a large loop gain makes the loop stay in lock with a virtually constant phase relationship between the input and VCO signals.

7.7 Phase Locked Loop Applications

Integrated circuit phase locked loops are versatile system building blocks suitable for use in a variety of frequency selective demodulation, signal conditioning or frequency synthesis applications. Practical considerations relating to phase locked loop operation have been discussed with reference to a specific device (the N E 565 general purpose phase locked loop) but our theoretical treatment of operating principles has been of a more general nature. The reader with a particular phase locked loop application in mind should be aware that a considerable range of integrated circuit phase lock loop devices are commercially available. A familiarity with the phase locked loop operating principles outlined in the previous sections should enable him more readily to appreciate information contained in the data sheets of specific devices.

Phase locked loop devices are available from several manufacturers : Signetics, Exar, Harris, Motorola, National, R.C.A., Texas. The upper operating frequency limit of currently available devices is of the order of 30 MHz. There are a whole range of general purpose phase lock loops and in addition there are specialised devices designed for specific applications. There are, for example, phase locked loop devices designed for sterio decoding, examples are: R.C.A. type CA 3090, Motorola type MC 1310P, Texas type SN 76115N, National LM 1310. There are devices intended for F.S.K. modulation and demodulation and tone decoding; examples are Signetics NE 562B, NE 567, XR 210 and XR 2567, a dual tone decoder. R.C.A. make a phase locked loop with very low power consumption, the type CD 4046A which takes a current of typically only 100 microamps from a 6 volt supply. This is of an order of 100 times less than the current required by most other monolithic loops.

Some of the basic ways in which phase locked loops have been applied will now be outlined, but like many other recently introduced integrated circuit devices their full range of application possibilities remains to be developed. When implementing a phase locked loop application, maximum effectiveness will be achieved if the user understands the factors controlling its performance. Having chosen a specific i.c. device, the user is in general free to select the centre frequency, lock range, and capture range and can tailor the dynamic response of the loop by his choice of loop filter; all of these factors have been discussed in the preceding sections.

7.7.1. F.M. Demodulation

When a phase locked loop is locked on a frequency modulated signal the VCO frequency tracks the instantaneous frequency of the input signal The filtered error voltage from the loop $V_{d(t)}$, which controls the VCO frequency, corresponds to the demodulated output. A practical investigation of the principle underlying this type of application was considered in section 7.5.3.

The linearity of the demodulated output is determined by the VCO voltage to frequency conversion characteristic. Phase locked loops can be used for detecting either wide band (high deviation) or narrow band F.M. signals with a higher degree of linearity that can be obtained with other F.M. detection systems. A phase locked loop functions as a self-contained receiver system since it combines the function of frequency selection and demodulation.

In frequency shift keyed (F.S.K.) data transmission systems, digital information is transmitted by switching the frequency of a signal between two defined values corresponding to a digital '1' and a digital '0'. A phase lock loop can be used for demodulation, the discrete changes in frequency resulting in discrete voltage steps in the loop error voltage.

7.7.2. A.M. Demodulation

The addition of a second phase sensitive detector (multiplier) to a phase locked loop system gives an arrangement for synchronous A.M. demodulation. Monolithic devices are available which provide this facility, for example Signetics type 561; their principle of operation is illustrated in the block diagram given in figure 7.21.

The phase locked loop locks on the carrier of the A.M. signal so that its VCO output has the same frequency as that of the carrier but no amplitude modulation. The demodulated A.M. is obtained by multiplying the VCO signal with the amplitude modulated input signal and filtering the output to remove all but the difference frequency component (see section 5.3.1). It will be remembered that when a phase locked loop is locked to an input signal at its centre frequency the VCO output signal is 90° out of phase with the input signal. The 90° phase shifting circuit is used in figure 7.21 so that the two

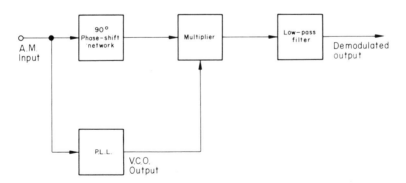

Fig. 7.21   Coherent amplitude modulation detection using a phase-locked loop

signals applied to the multiplier shall be in phase. The average value (difference frequency component) of the multiplier output is then directly proportional to the amplitude of the input signal. Phase locked A.M. detection is a coherent detecting technique, it exhibits a high degree of selectivity centred about $f_0{}^1$ and offers a higher degree of noise immunity than conventional peak detector type A.M. demodulators.

7.7.3. Phase Modulation

In a phase locked loop which is locked onto an input signal at the centre frequency of the loop there is a 90° phase difference between the VCO signal and the input signal. If a current is injected into the loop output terminal (the low pass filter output) the phase of the VCO signal shifts in order that the phase detector may develop an opposing average current. Thus the VCO control voltage may remain constant and maintain the lock.

A block diagram illustrating this type of application is shown in figure 7.22. If the input signal is a square wave the phase of the VCO signal will be a linear function of the injected current.

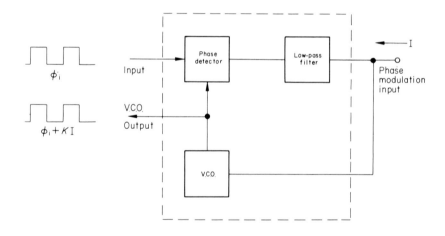

Fig. 7.22   Phase modulation using the PLL

7.7.4. Frequency Synthesis

Phase locked loops can be used to generate signals with frequency precisely related (by an exact multiple) to that of a reference signal. Frequency multiplication can be achieved in two different ways: by arranging that the loop shall lock to a harmonic of the reference signal or by breaking the loop and inserting a digital frequency divider (a counter) so that the loop locks to the divided VCO frequency. The two techniques can be combined in order to generate a frequency which is related to that of the reference signal by some exact fractional relationship.

Harmonic locking is the simplest and is achieved by setting the VCO free running frequency to a multiple of the input frequency. The VCO frequency then locks to the required harmonic of the input frequency. The limitation to this scheme is that the lock range decreases as successively higher and weaker harmonics are used for locking. This limits the practical harmonic range to multiples something less than 10.

The second scheme for frequency multiplication is illustrated by the block diagram shown in figure 7.23. The loop is broken and a counter inserted,

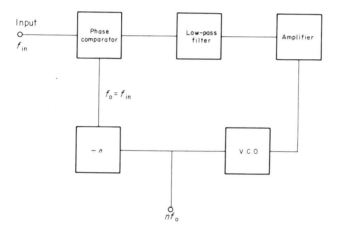

Fig. 7.23 Frequency synthesiser using frequency divider

between the VCO and the phase detector. The amount of frequency multiplication is directly determined by the countdown number of the counter. If a fractional relationship is required, say 8/3, the VCO centre frequency is initially set to approximately $8/3\, f_{in}$. The counter is set to divide the VCO frequency by 8 and the input signal locks to the third harmonic of

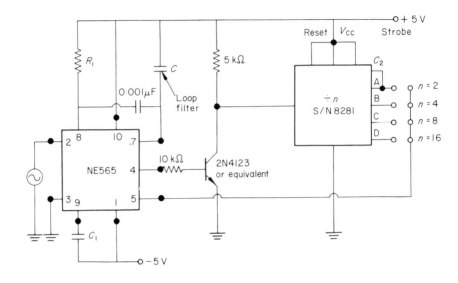

Fig. 7.24   Implementation of frequency synthesiser

the VCO signal. A practical circuit for frequency multiplication using the NE 565 phase locked loop is shown in figure 7.24. Note that in frequency multiplication applications it is usually necessary to use a loop filter with quite a large time constant in order to remove the sum component in the phase detector output and avoid an incidental frequency modulation of the VCO signal.

7.7.5 Frequency Translation

A phase locked loop can be used to translate, or 'offset', the frequency of a fixed frequency reference oscillator by a small amount. Frequency offset is achieved by adding a mixer (multiplier) and a low pass filter to the basic loop as shown by the block diagram in figure 7.25.

The reference signal and the VCO signal are applied as input signals to the multiplier which forms the sum and difference components. The low pass filter removes the sum component and the different component, $f_0 - f_r$ is applied as an input to the loop phase detector. A signal at the required offset frequency f_1 is used as the second input to the loop phase detector and when the loop goes into lock the two signals applied to the phase detector are held at the

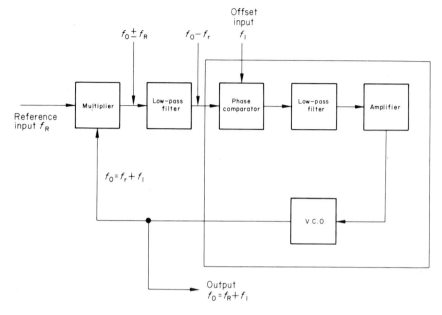

Fig. 7.25   Frequency translation or offset loop

same frequency. Thus

$$f_0 - f_r = f_1$$
and
$$f_0 = f_r + f_1$$

The VCO signal frequency is that of the reference signal offset by the amount f_1.

It is hoped that the somewhat brief outline of phase locked loop applications that has been given may have suggested application areas of interest to the reader. Details of further applications and specific circuits are to be found in Reference 5.

References.

1.   F.M. Gardner. *Phaselock Techniques,* John Wiley & Sons, New York, (1966)

2.   A.J. Viterbi. *Principles of Coherent Communication,* McGraw-Hill, New York,(1966)

3. J. Klappes and J.T. Frankle. *Phase-Locked and Frequency Feedback Systems,* Academic Press, New York and London, (1972)

4. G.S. Moschytz. Miniaturised RC filters using phase locked loop *Bell System Tech. J.,* **44** (May 1965), 823-70

5. *Linear Phase-Locked Loops Applications Book,* Signetics.

Exercises 7

7.1 A four quadrant linear multiplier (scale factor 1/10), a simple first order low pass filter (time constant = 2×10^{-2} seconds) and a VCO are connected together to form a phase locked loop. Sketch the arrangement. The VCO produces a square wave output signal, peak to peak value 20 volts, symmetrical about earth. Its centre frequency is 10^3 radians per second and its conversion gain is 500 radians per second per volt. Find the fundamental lock range and the approximate capture range for a sinusoidal input signal of amplitude 2 volts. What is the maximum d.c. error voltage generated by the loop? Sketch the transfer characteristic between input signal frequency and d.c. error voltage. Find the lock range for a sinusoidal signal amplitude 2 volts with frequency varied about 3 times the VCO centre frequency.

7.2 If the phase locked loop of question 7.1 is locked to a sinusoidal input signal of angular frequency 1,200 radians per second and amplitude 2 volts, what is the phase difference between the VCO signal and the input signal? What happens to this phase difference if the amplitude of the input signal is reduced to 1 volt? What is the lock range and approximate capture range for a signal of amplitude 1 volt?

7.3 From the loop transfer curves given in figure 7.7 deduce the conversion gain of the VCO used in the loop and the approximate magnitude of the low pass filter resistor.
Note: Equation 7.10 may be written as:

$$\omega_c \cong \left(\frac{\omega_L}{\tau}\right)^{1/2}$$

The narrower the capture range the better the approximation.

7.4 A phase locked loop has the following performance characteristics:

$f_0' = 250$ kHz , $K_0 = 500 \times 10^3$ rad/s/volt.

$K_d = 0.45$ volt/radian , $A = 1.4$

What is the lock range of the loop? A frequency modulated input signal of the form

$$V_{s(t)} = V_s \cos\left[520 \times 10^3 \pi + \frac{250 \times 10^3}{\omega_m} \sin \omega_m t\right]$$

is applied to the loop. What is the constant phase error ϕ_K generated by the loop? What is the amplitude of the demodulated output signal produced by the loop? At what modulation frequency is the demodulated output signal 3dB down on its low frequency value? No loop filter is used. Assume that filtering is performed outside the loop.

7.5 A simple first order low pass filter, time constant 2×10^{-5} seconds is included inside the phase locked loop of question 7.4. Discuss the effect that this has on the performance of the loop, sketch the appropriate Bode plots. What is the damping factor associated with the second order loop so formed? How much peaking do you expect in the magnitude/frequency response of the demodulated output signal? If instead of the sinusoidal frequency modulation of the input signal, the frequency is modulated by a low frequency squarewave, sketch the waveform that you expect the demodulated output signal to have. What is the settling time of the demodulated output signal to within 1 per cent accuracy?

7.6 A phase locked loop has the following characteristics:

$f_0 = 1$ MHz $K_0 = 2 \times 10^6$ rad/s/volt,

$K_d = 0.45$ volts/radian, $A = 1.4$

An input signal frequency modulated about the loop's centre frequency is applied, find the value of τ_1 and τ_2 to be used in a lag lead type of filter to give the loop a response with damping factor $1/\sqrt{2}$ and natural frequency 20 kHz (see figure 7.14). Sketch the appropriate Bode plots.

7.7 Derive an expression for the VCO conversion gain given by the arrangement for increasing lock range shown in figure 7.20.

7.8 An amplitude modulated signal with carrier frequency equal to the phase locked loop centre frequency is applied to the amplitude modulation detection system illustrated in figure 7.21. Consider the various frequency components present in the multiplier input and output signals.

7.9 Analyse the action of the phase modulation scheme outlined in figure 7.22.

7.10 Sketch an arrangement for producing a signal with frequency precisely 10/3 of the frequency of a reference signal.

Appendix

Input Output Relationships – Transfer Function Terminology

The mathematical techniques of network analysis allow the precise formulation of equations which express the response of linear networks, (linear amplifier circuits, filters etc) to a variety of input signals. The practising engineer may have neither the time nor inclination to pursue network theory in depth but it is desirable that he should have a sufficient acquaintance with its concepts and terminology to enable him to make practical use of the equations developed to describe linear network behaviour. This section is designed to provide such a working knowledge, those requiring a greater understanding of the concepts involved, are referred to one of the many books on the subject.[1,2]

Consider an electrical network consisting of linear elements with an input signal $V_{i(t)}$, the network responds to produce an output $V_{o(t)}$. The action of the network on the input signal to produce the output signal can be expressed by a linear differential equation of the form

$$a_m \frac{d^m}{dt^m} V_{i(t)} + a_{m-1} \frac{d^{m-1}}{dt^{m-1}} V_{i(t)} + \ldots + a_0 V_{i(t)}$$

$$= b_n \frac{d^n}{dt^n} V_{o(t)} + b_{n-1} \frac{d^{n-1}}{dt^{n-1}} V_{o(t)} + \ldots + b_0 V_{o(t)} \quad (A.1)$$

a_m and b_n are constant coefficients determined by the values of the network elements, the largest value of m or n specifies the order of the network. Symbolically we may write:

$$p = \frac{d}{dt}, \quad p^2 = \frac{d^2}{dt^2}, \text{ etc. when equation 1 becomes}$$

$$a_m p^m V_{i(t)} + a_{m-1} p^{m-1} V_{i(t)} + \ldots + a_0 V_{o(t)}$$

$$b_n p^n V_{o(t)} + b_{n-1} p^{n-1} V_{o(t)} + \ldots + b_0 V_{o(t)}$$

Linear Integrated Circuit Applications

and the ratio of output to input is

$$\frac{V_{o(t)}}{V_{i(t)}} = \frac{a_m p^m + a_{m-1} p^{m-1} + \ldots + a_0}{b_n p^n + b_{n-1} p^n + \ldots + b_0} = A_{(p)} \quad (A.2)$$

Equation A.2 is a symbolic representation of the differential equation, $A_{(p)}$ is called the transfer operator of the network.

If the input signal is a sine wave function of time, $V_{i(t)} = v_i \sin \omega t$ then

$$p V_{i(t)} = \omega V \cos \omega t = j\omega\, v_i \sin \omega t = j\omega\, V_{i(t)}$$

and

$$p^2 V_{i(t)} = -\omega^2 v_i \sin \omega t = -\omega^2 V_{i(t)}$$

An expression for the sinusoidal response of the network *may* be obtained by substituting $p = j\omega$, $p^2 = -\omega^2$, $p^3 = -j\omega^3$, $p^4 = \omega^4$ ... etc. in equation A.2.

The sinusoidal response function is a special case of the more general transfer function which is obtained by considering the operator p as a complex variable $s = \sigma + j\omega$, ($\sigma = 0$ in the case of the sinusoidal response).

The transformation from the differential equation in the time domain to the complex frequency domain (the s domain) may be effected by considering the input signal to be an exponential function of the form $V_i = V e^{st}$, the output signal will then be an exponential function with the same time response characteristics. The choice of an exponential signal rests in the fact that a large variety of signals can be considered as made up of appropriately chosen exponential functions.

If $\quad V_{i(t)} = V_i e^{st}$

$$p V_{i(t)} = \frac{d}{dt} V_i e^{st} = s V_{i(t)}$$

and

$$p^2 V_{i(t)} = s^2 V_{i(t)} \text{ etc}$$

We replace p by the complex variable s in equation A.2 and at the same time, factor the polynominal numerator and denominator so as to express the equation as a ratio of the product of the roots. The equation for the transfer function is then written as

$$A_{(s)} = \frac{a_m}{b_n} \frac{(s-Z_m)}{(s-P_n)} \frac{(s-Z_{m-1})}{(s-P_{n-1})} \ldots \frac{(s-Z_1)}{(s-P_1)}$$

or

$$A_{(s)} = a_0 \frac{(1-\frac{s}{Z_m})(1-\frac{s}{Z_{m-1}}) \cdots (1-\frac{s}{Z_1})}{b_0 (1-\frac{s}{P_n})(1-\frac{s}{P_{n-1}}) \cdots (1-\frac{s}{P_1})} \quad (A.3)$$

Those values of the complex variable, s, which cause the denominator of the transfer function to be zero are called poles and those values of s which cause the numerator to be zero are called zeroes. Necessary and sufficient conditions for a transfer function to be physically realisable and stable are: the real parts of all poles should be negative, the gain at zero frequency a_0/b_0, should be finite and the order of the numerator should not exceed that of the denominator. Provided that numerator and denominator polynomials have all real coefficients it can be shown that all complex poles and zeros must occur in conjugate pairs ($\sigma \pm j\omega$).

We will be most concerned with first and second order network transfer functions and we now consider the nature of these functions in greater detail.

Low-Pass Transfer Functions

1st Order A general expression for a first order (single pole) low pass function may be written in terms of the complex frequency variable s as

$$A_{(s)} = \frac{A_0}{1 + \frac{s}{\omega_0}} \quad (A.4)$$

The response to steady state sinusoidal excitation is

$$A_{(j\omega)} = \frac{A_0}{1 + j\frac{\omega}{\omega_0}}$$

with magnitude

$$|A_{(j\omega)}| = \frac{A_0}{\left(1 + \left(\frac{\omega}{\omega_0}\right)^2\right)^{1/2}} \quad (A.5)$$

and phase angle

$$\theta = \tan^{-1}-\frac{\omega}{\omega_0} \quad (A.6)$$

As an example of a network which exhibits a first order characteristic a simple resistor capacitor low pass filter has a transfer function

$$\frac{V_{0(s)}}{V_{i(s)}} = F_{(s)} = \frac{1}{1 + CR\ S}$$

$\omega_0 = \dfrac{1}{CR}$ is the so-called 'break frequency' of the circuit.

2nd Order A general expression for a second order low pass function may be written as

$$A_{(s)} = \dfrac{A_0}{1 + b\dfrac{s}{\omega_0} + \dfrac{s^2}{\omega_0^2}} \qquad (A.7)$$

The poles of the response, obtained as the roots of the denominator are

$$p_{1,2} = \dfrac{-\dfrac{b}{\omega_0} \pm \left(\dfrac{b^2}{\omega_0^2} - \dfrac{4}{\omega_0^2}\right)^{1/2}}{\dfrac{2}{\omega_0^2}}$$

$$= \dfrac{-b}{2}\omega_0 \pm \omega_0 \left(\left(\dfrac{b}{2}\right)^2 - 1\right)^{1/2} \qquad (A.8)$$

It is usual to write $\dfrac{b}{2}$ as ζ the so-called damping factor.

If $\zeta < 1$ the response has a pair of complex conjugate poles

$$p_{1,2} = -\zeta\omega_0 \pm j\omega_0 (1 - \zeta^2)^{1/2}$$

If $\zeta \geq 1$ the response has two real poles and can be considered as the product of two first order functions.

The magnitude of the response for steady state sinusoidal signals is

$$|A(j\omega)| = \dfrac{A_0}{\left(\left[1 - \left(\dfrac{\omega}{\omega_0}\right)^2\right]^2 - 4\left(\zeta\dfrac{\omega}{\omega_0}\right)^2\right)^{1/2}} \qquad (A.9)$$

The phase angle is

$$\theta = \tan^{-1} \dfrac{-2\zeta\dfrac{\omega}{\omega_0}}{1 - \left(\dfrac{\omega}{\omega_0}\right)^2} \qquad (A.10)$$

Neglecting A_0 as a scaling factor, plots of the magnitude and phase response for different values of the damping factor are shown in figure A.1.

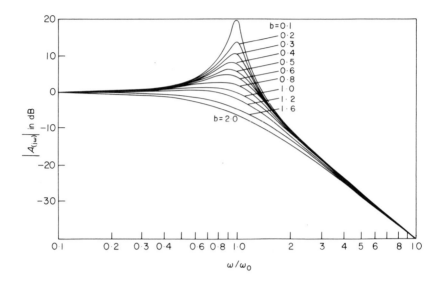

Fig. A.1 Magnitude response of second order, low-pass transfer function for several values of b ⩽ 2

Note that the magnitude shows increasing amounts of peaking as b is reduced. The angular frequency at which peaking occurs may be found by differentiating equation A.9 with respect to ω and equating to zero. This gives the peaking frequency

$$\omega_p = \omega_0 \, (1 - 2\zeta^2)^{1/2} \tag{A.11}$$

$$\zeta < \frac{1}{\sqrt{2}}$$

Substituting this value of ω in equation A.9 gives the peak value of $|A_{j\omega}|$ as

$$|A_{j\omega}|_{\max} = \frac{A_0}{2\zeta(1-\zeta^2)^{1/2}} \tag{A.12}$$

The extent of the response magnitude peaking may be written as

$$P_{\text{(dB of peaking)}} = 20 \log_{10} \frac{1}{2\zeta(1-\zeta^2)^{1/2}}$$

$$\zeta < \frac{1}{\sqrt{2}} \tag{A.13}$$

Linear Integrated Circuit Applications

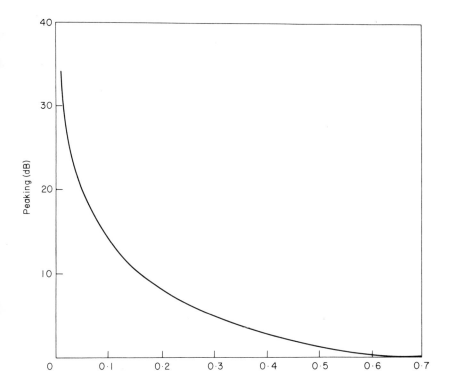

Fig. A.2.   Peaking (dB) as a function of ζ

p represents the increase in magnitude of the response above A_0 because of peaking. A knowledge of the damping factor for a second order response enables the magnitude of the frequency response peaking to be calculated. Figure A.2 is a plot of equation A.13 which shows graphically the effect of a ζ on peaking.

In a practical evaluation of a second order low pass system a measurement of the response of the system to a voltage step input often provides a rapid means of characterising the system. A system with the transfer function equation A.7 when supplied with a step input E_i gives an output signal which is described by the equation.[3]

$$e_{0(t)} = A_0 E_i \left[1 - \frac{e^{-\zeta \omega_0 t}}{(1-\zeta^2)^{\frac{1}{2}}} \sin\left(\omega_0 \sqrt{1-\zeta^2}\, t + \cos^{-1}\zeta\right) \right] \quad (A.14)$$

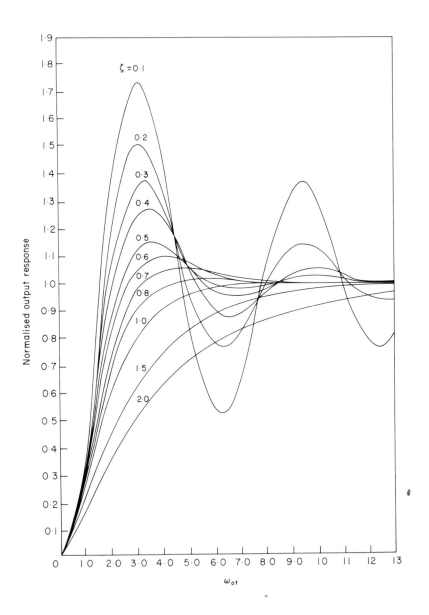

Fig. A.3   Second order step response

Linear Integrated Circuit Applications 255

The transient response is seen to be directly determined by the constants ζ and ω_0 of the transfer function.

Treating $A_0 E_i$ as a scaling factor and plotting time in units of $\omega_0 t$ the normalised step response for different values of ζ is shown plotted in figure A.3. The step response shows an increasing overshoot and ringing as the value of the damping factor is successively reduced below unity. The ringing frequency is defined by the sine term in equation A.14 and is a damped frequency of oscillation represented by the equation

$$\omega_d = \omega_0 \cdot (1 - \zeta^2)^{1/2} \tag{A.15}$$

Note that ω_0 represents the natural (or undamped) frequency of oscillation obtained when the damping factor is zero.

The times at which the peaks of the ringing occur can be found by differentiating equation A.14 with respect to time and equating to zero, this operation gives

$$t = \frac{N\pi}{\omega_0 (1 - \zeta^2)^{1/2}} \qquad N = 1, 2, 3, \tag{A.16}$$

Substituting these values of t back into equation A.14 gives

$$\left| \frac{e_{0(t)}}{A_0 E_i} \right|_{\text{1st max.}} = 1 + e^{\frac{-\zeta\pi}{(1-\zeta^2)^{1/2}}} = V_{p_1}$$

$$\left| \frac{e_{0(t)}}{A_0 E_i} \right|_{\text{1st min.}} = 1 - e^{\frac{-2\zeta\pi}{(1-\zeta^2)^{1/2}}} = V_{p_2}$$

$$\left| \frac{e_{0(A)}}{A_0 E_i} \right|_{\text{2nd max.}} = 1 + e^{\frac{-3\zeta\pi}{(1-\zeta^2)^{1/2}}} = V_{p_3} \tag{A.17}$$

The amount by which $\left|\frac{e_{0(t)}}{A_0 E_i}\right|_{\text{1st max.}}$ exceeds unity is called the overshoot, which, expressed as a percentage is

$$\text{Overshoot \%} = 100\% \; e^{\frac{-\zeta\pi}{(1-\zeta^2)^{1/2}}} \tag{A.18}$$

Overshoot represents the maximum output transient error following an initial rise in response to a step input. The time taken by the output to settle within a certain accuracy (settling time) following a transient, is often of greater interest. A conservative estimate of the small signal settling time for a second order low pass function can be made by finding the smallest value of N which satisfies

$$100\% \; \frac{e^{-\zeta N \pi}}{(1-\zeta^2)^{1/2}} \leqslant x\%$$

where $x\%$ represents a specified accuracy. This value of N is then substituted into equation A.16 to give the estimate of settling time.

A measurement of the first three ringing peaks and the times at which they occur can be used to evaluate ζ and ω_0 for an underdamped second order system. For using equations A.17 we may write

$$\frac{Vp_1 - Vp_3}{Vp_2} = \frac{1 + e^{\frac{-\zeta\pi}{(1-\zeta^2)^{1/2}}} - 1 - e^{\frac{-3\zeta\pi}{(1-\zeta^2)^{1/2}}}}{1 - e^{\frac{-2\zeta\pi}{(1-\zeta^2)^{1/2}}}}$$

$$= e^{\frac{-\zeta\pi}{1-\zeta^2}}$$

Taking logarithms

$$\log_{10}\left(\frac{Vp_1 - Vp_3}{Vp_2}\right) = \frac{-\zeta\pi}{2.3\,(1-\zeta^2)^{1/2}}$$

Solving for ζ

$$\zeta = \frac{\log_{10}\frac{Vp_1 - Vp_3}{Vp_2}}{\left(\log_{10}^2 \frac{Vp_1 - Vp_3}{Vp_2} + 1.8615\right)^{1/2}} \tag{A.19}$$

Using equation A.16 we may write

$$\omega_0 = \frac{2\pi}{(t_3 - t_1)\,(1-\zeta^2)^{1/2}} \tag{A.20}$$

where t_3 and t_1 are the times at which the 1st and 3rd peaks occur. Having calculated ζ from equation A.19, equation A.20 can be used to find ω_0.

Many practical systems which are of interest are characterised by a second order low pass transfer function. This gives the mathematical relationships that we have discussed considerable value when dealing quantitatively with the behaviour of such systems. High pass and band pass functions which we now discuss show some similarities with the low pass functions.

High Pass Transfer Functions

1st Order A general expression for a first order high pass function may be written as

$$A_{(s)} = \frac{A_0 s}{s + \omega_0} \tag{A.21}$$

The response has a single zero equal to nothing, (a time independent signal, i.e. d.c., gives zero output), and a single real pole.

The response to a steady state sinusoidal input signal is

$$A_{(j\omega)} = \frac{A_0}{1 - j \frac{\omega_0}{\omega}}$$

with magnitude

$$|A_{(j\omega)}| = \frac{A_0}{\left(1 + \left(\frac{\omega_0}{\omega}\right)^2\right)^{1/2}}$$

and phase angle

$$\theta = \tan^{-1} \frac{\omega_0}{\omega}$$

2nd Order A general expression for a second order high pass function may be written as

$$A_{(s)} = \frac{A_0 s^2}{s^2 + b s \omega_0 + \omega_0^2} \tag{A.22}$$

The response has two zeroes both equal to nothing and two poles. If $b/2 = \zeta < 1$ the poles are a complex conjugate pair. If $\zeta \geq 1$ the poles are real and the response can be considered as the product of two 1st order high pass functions.

The magnitude of the response for a steady state sinusoidal input signal is

$$|A_{(j\omega)}| = \frac{A_0 \omega^2}{(\omega^4 + \omega_0^4 + \omega^2 \omega_0^2 (b^2 - 2))^{1/2}} \qquad (A.23)$$

and the phase angle

$$\theta = \tan^{-1} \frac{b \omega_0 \omega}{\omega^2 - \omega_0^2} \qquad (A.24)$$

Neglecting A_0 as a scaling factor the magnitude response for various values of $b/2 = \zeta$ is shown graphically in figure A.4.

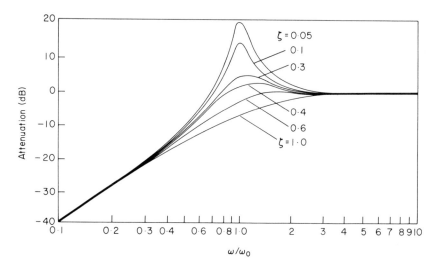

Fig. A.4 $|A_{jw}|$ in dB against ω/ω_0 for different values of ζ for two-zero, two-pole transfer function

Band Pass Transfer Function

A general expression for a second order band pass transfer function may be written as

$$A_{(s)} = \frac{A_0 b \omega_0 s}{s^2 + b\omega_0 s + \omega_0^2} \qquad (A.25)$$

The response has a single zero equal to nothing and two poles. Examination of the equation shows that if $\zeta < 1$ the poles are a complex conjugate pair; $\zeta = b/2$, the damping factor as used previously. If $\zeta \geqslant 1$ the poles are real and the response may be regarded as a product of a 1st order low pass function and a first order high pass function.

We consider the underdamped case, $\zeta < 1$, and write

$$2\zeta = b = \frac{1}{Q}$$

$$\text{where } Q = \frac{\dot{\omega}_0}{\omega_2 - \omega_1} = \frac{f_0}{f_2 - f_1} \quad \text{(A.26)}$$

f_2 and f_1 are the frequencies at which the magnitude response is 3 dB down on A_0. The response for a steady state sinusoidal input signal may then be

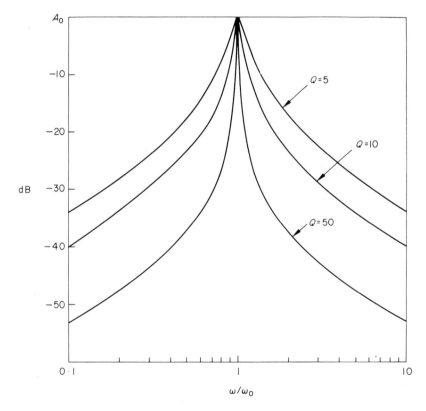

Fig. A.5   Second order band pass transfer function

written as

$$A_{(j\omega)} = \frac{A_0}{1 + jQ\left[\dfrac{\omega}{\omega_0} - \dfrac{\omega_0}{\omega}\right]} \qquad (A.27)$$

with magnitude

$$|A_{(j\omega)}| = \frac{A_0}{\sqrt{1 + Q^2\left[\dfrac{\omega}{\omega_0} - \dfrac{\omega_0}{\omega}\right]^2}} \qquad (A.28)$$

and phase angle

$$\theta = \tan^{-1} -Q\left[\dfrac{\omega}{\omega_0} - \dfrac{\omega_0}{\omega}\right] \qquad (A.29)$$

The magnitude of the response for several values of Q is shown graphically in figure A.5.

References

1. F. Kuo Franklin. *Network Analysis and Synthesis,* John Wiley & Sons New York, (1962)
2. M.E. Van Valkenberg. *Network Analysis,* Prentice Hall, New Jersey, (1964)
3. J.J. DiSteffano, A.R. Stubberd and I.J. Williams. *Feedback and Control Systems,* McGraw-Hill, New York, (1967)

Answers to Exercises

Exercise 1

1.2 $\text{CMRR} = \dfrac{1 + A_{\text{CL}}}{4K}$

1.3 $R_1 = 50 \text{ k}\Omega$, $R_2 = 500 \text{ k}\Omega$ 16 mV

1.6 Circuit of figure 1.5 with $R_1 = 20 \text{ k}\Omega$, $R_2 = 200 \text{ k}\Omega$. Input Resistance = 10 kΩ

1.7 69 dB   76 dB

1.8 (a) $R = 2 \text{k}\Omega$ $R_1 = 10 \text{k}\Omega$ $R_2 = 100 \text{ k}\Omega$
    (b) 71 dB

1.9 59 dB

1.10 $\alpha \geqslant 0.006$

1.11 $R = 100 \: \Omega$ 1%

1.12 $\dfrac{R_2}{R_1} = 99$ 1%

Exercise 2

2.1 $R_1 = 50 \text{ k}\Omega$ $R_2 = 500 \text{ k}\Omega$ $C = 0.32 \: \mu\text{F}$
    27 mv

2.2 $R_1 = 4.39 \text{ k}\Omega$, $R_2 = 57.7 \text{ k}\Omega$ $\dfrac{R_4}{R_3} = 9$

2.3 $R_1 = R_2 = 2.25 \text{ M}\Omega$
    (a) 2.25 V
    (b) 45 mV

2.4 $R = 22.5 \text{ k}\Omega$ $\dfrac{R_4}{R_5} = 3.72$ 24.4 dB

2.5 Low Pass $A_{(j\omega)} = \dfrac{1.82}{1 + j\,0.182 \left[\dfrac{\omega}{10^4} - \dfrac{\omega^2}{10^8}\right]}$

Band Pass $\quad A_{(j\omega)} = -\dfrac{10}{1 + j\,5.5\left[\dfrac{\omega}{10^4} - \dfrac{10^4}{\omega}\right]}$

High Pass $\quad A_{(j\omega)} = -\dfrac{1.82\,\omega^2}{10^8 + j\,1.82 \times 10^3\,\omega - \omega^2}$

2.6 $\quad A_{(j\omega)} = \dfrac{10^6 - \omega^2}{10^6 + j\,4 \times 10^3\,\omega - \omega^2}$

$\quad A_{(j\omega)} = \dfrac{10^6 - \omega^2}{10^6 + j\,2 \times 10^2\,\omega - \omega^2}$

$\quad\quad$ 0.012 V $\quad\quad$ 0.102 V

2.7 $\quad \zeta = 0.2 \quad \omega_0 = 10^3$ rad/s

$\quad\quad \tau_s = 35.2$ ms $\quad P_{dB} = 8.14$ dB

Exercise 3

3.1 $\quad T_1 = 0.21$ ms, $\quad T_2 = 0.14$ ms

3.2 $\quad f = 4.76$ kHz, $\quad R_2 = 38$ kΩ

3.3 $\quad$ 16.7 kHz

3.4 $\quad t_1 = 1.7 \times 10^{-4}$ seconds $\quad t_2 = 2.5 \times 10^{-4}$ seconds

Exercise 4

4.1 $\quad$ 3.5 volts, $\quad$ 2.875 volts and 4.125 volts

4.2 $\quad V = 77\,V_x\,V_y$

Exercise 5

5.2 $\quad$ 10.1 volts

5.3 $\quad$ 0.84 volts

Exercise 6

6.1 $\quad$ 0.52%

6.2 $\quad$ 5.5% $\quad\quad$ 3.3%

6.3 $\quad K = 20 \quad R_x = 3.3$ kΩ $\quad R_y = 3.3$ kΩ $\quad R_L = 54$ kΩ

6.5 $\quad R = 10$ kΩ $\quad R_A = 5.83$ kΩ $\quad R_B = 11.7$ kΩ
$\quad\quad R_C = 17.5$ kΩ $\quad R_D = 23.3$ kΩ

Exercise 7

7.1 $\omega_L = 637$ rad/s $\quad \omega_C = 178$ rad/s $\quad 1.27$ V $\quad \omega_L = 212$ rad/s

7.2 $71.7°$ $\quad 51°$ $\quad \omega_L = 318$ rad/s $\quad \omega_C = 126$ rad/s

7.3 37.7×10^3 rad/s/volt 5 kΩ 2.6 kΩ

7.4 50 kHz $\quad \phi_K = 11.5°, \ 0.5$ V $\quad f_{3dB} = 49$ kHz

7.5 $\zeta = 0.2$ $\quad P_{dB} = 8$dB $\quad \tau_s = 2.1 \times 10^{-4}$ s

7.6 $\tau_1 = 6.93 \times 10^{-5}$ s $\quad \tau_2 = 1.05 \times 10^{-5}$ s

7.7 $K_0 = \dfrac{5\pi R_f}{V_s R_B R_1 C_1}$

Index

A.C. measurement, 44, 45
Active filters, 52–88
    analogue computer realisation, 69–73
    band pass, 62–68
    band reject, 73–80
    design tables, 87
    first order high pass, 59
    first order low pass, 53
    second order high pass, 60–63
    second order low pass, 54
    tuning, 56
    V C V S circuits, 54–64
Adder subtractor, 69
AM demodulation, 241
Amplitude modulation (AM), 151, 155, 162, 163
Analogue computing techniques, 69, 70
Analogue switch, 152
Asymmetrical waveforms, 107, 110, 115
Automatic level control, 191–194
Averaging filter, 185

Balanced modulator, 128, 131, 133, 151, 155, 158, 161, 162
Band pass active filters, 62–68, 71
    design procedure, 63–65, 67, 68, 72
Band reject filters, 73–80
    active twin T, 77, 78
    derived from band pass, 75
    passive twin T, 76
    rejection bandwidth, 78
    transient response, 80
Bessel filter, 82, 83, 85
Bias current, 48, 148, 166
Biasing, 158, 159, 215
Bode plot, 228–230, 232
Bridge amplifiers, 24–30
    choice of, 25
    current read out, 27
    differential input, 25
    earthed unknown, 27
    half bridge, 28, 29
    linear read out, 26
    passive bridge, 24
    single ended, 28, 29

Cable capacitance, 37
Capacitive microphone, 35, 36
Capacitive transducer, 34, 35
Capture process, 212
Captive range, 212, 213
Capture transient, 213
Centre frequency, 210, 213
Characteristic time, 69
Charge amplifier, 32–37
Closed loop gain, 46
Common mode input resistance, 3
Common mode range, 7
Common mode rejection ratio (CMRR), 2–5, 8, 21, 22

266 *Index*

Comparator, 92
Complex variable, 249, 250
Computation circuits, 194, 196–198
Computing loop, 69
Controlled gain devices, 125–141
Controlled operational amplifiers, 126, 127, 143–147
Conversion gain, 221
Current booster, 11, 16
Current integrator, 32, 34, 35
Current inversion negative inmittance converter (INIC), 65–68
Current measurement, 41, 43, 44
Current source, 11, 16, 17
Current to voltage conversion, 30, 41

Damping factor, 54, 229, 232, 251, 253
Data amplifier, *see* Instrumentation amplifier
Demodulation, 156, 233, 240
Difference of squares, 196
Differential equations, 248, 249
Differential input amplifier
   high input impedance, 5, 6
   one amplifier circuit, 2
   with large common mode range, 7
   with variable gain, 4, 6
Differential input measurements, 17–23
Differential input resistance, 3
Differential output, 151
Differential output amplifier, 7–9
Differential output current, 122, 134, 135
Dividing, 182, 183, 191
Dosimetry, 32

Earth current, 18

Earth loop, 18
Errors in multiplier, 165–173

Feedback, 12, 43, 46
Feedback fraction, 46
Feedthrough, 168, 169
FET input amplifier, 3, 39, 43
Filtering by frequency changing, 203
Filters, *see* Active filters
First order low pass filter, 53
Flow control, 194
FM demodulation, 240
Frequency changing, 203
Frequency compensation of i.c. instrumentation amplifier, 14
Frequency doubling, 156, 181
Frequency modulated signal, 234
Frequency modulation (FM), 109, 111, 116–118
Frequency multiplication, 242–244
Frequency offset, 244, 245
Frequency response of multiplier, 170, 176, 179
Frequency shift keyed (FSK), 240–241
Frequency spectra, 162, 163, 205, 206
Frequency sweep, 116, 218
Frequency synthesis, 242–244
Frequency translation, 244, 245
Functional relationships, 198, 199

Gain
   control, 125, 128
   error, 47
   setting, 10, 11, 13, 14
   variable, 131, 143–163
Gate control characteristic, 150
Gate controlled amplifier, 128
Gated generator, 105, 106

Index

Gated oscillator, 153, 154

Half bridge, 28, 29
High pass filters, 59–63, 72
   design procedure, 61, 62, 72
   tuning, 61
Hum, 18

Input bias current, 48, 148, 149, 166, 177
Input offset current, 166
Input offset voltage, 48, 148, 166, 177
Instrumentation amplifiers, 1–19
   gain setting, 10, 11, 13, 14
   integrated circuit, 9–14
   offsetting output, 15, 16
   setting up procedure, 13, 14
   use of sense and reference terminals, 14–17, 19

Lag/lead filter, 230, 231
Level control 191–194
Loop gain, 46
Loudspeaker power, 190
Low pass filters, 53–59, 71
   component choice, 55
   design procedure, 55–58

Mains hum, 18
Mean square, 185
Measurement amplifiers, *see* Instrumentation amplifiers
Meter circuits, 39–45
Meters, pointer-scale, 39, 40
Modulation
   amplitude, 151, 155, 158 162
   balanced, 128, 130–133, 151, 155, 158, 160
   circuits, 157–163
   linear, 161, 202
   nonlinear, 161, 202

phase, 242
processes, 155–157
pulse position, 95
pulse width, 99
Modulator applications, 155–163, 202–206
Monostable, 96, 100
Multiplexer, 146, 147, 152, 153
Multipliers, 136–141, 165–206
   applications, 180–206
   errors, 165–173
   feedthrough, 168, 169
   four quadrant linear, 165–206
   frequency response, 170–179
   functional schematic, 167
   ideal performance, 165
   nonlinearity, 156, 171, 172, 178, 179
   offsets, 168, 169, 177, 181
   scale factor error, 168, 169
   scaling factor, 139, 165, 169, 176
   slewing rate, 180
   specifications, 165, 166, 168
   test circuits, 173–179
   total d.c. error, 173
   transfer curve, 171
   variable scaling factor, 140, 141, 186, 196, 197
   vector error, 171
Multipole filters, 81–88
Multivibrator, 91

Natural frequency, 54, 56, 61, 229, 232, 255
Negative inmittance converter, 65–68
Negative ramp, 107, 109
Noise gain, 47
Nonlinearities, 156, 171, 172, 178, 179
Nonlinear processing, 198, 199
Notch filter, *see* Band reject

Index

Offsets, 47, 48, 168, 169, 177, 181
Open loop gain, 46
Oscillators
   amplitude stability, 154
   gated, 153, 154
   quadrature, 194, 195
   stabilisation, 193
   variable duty cycle, 101
   Wien bridge, 154, 193
Output offset, 15, 16
Overshoot, 255, 256

Peaking, 252, 253
Phase error, 223, 225
Phase locked loop (PLL), 208–245
   applications, 239–245
   block diagram, 210
   building blocks, 208, 209
   capture characteristics, 222
   capture process, 212
   capture range, 212, 214, 219, 221, 224
   capture transient, 213
   centre frequency, 209
   circuit schematic, 217
   damping, 229–234
   dynamic behaviour, 224–235
   first order loop, 227, 228
   hold-in range, 213
   lock range, 210, 213, 214, 219, 221, 223
   lock range increase, 238, 239
   lock range reduction, 236–238
   loop filter, 215, 222, 227, 229–231
   measurement loop transfer curve, 214–219
   phase amplifier representation, 226
   phase detector conversion gain, 221
   phase error, 223, 225
   pull-in range, 213
   pull-in time, 212
   second order loop, 228–233
   test circuit, 216, 233
   tracking, 210
   tracking range, 213, 223
   transient response, 229, 233–235
Phase modulation, 242
Phase sensitive detector, 156, 202, 203, 208–211, 221
Photocell amplifier, 30–32
Pick up, 18
Piezoelectric transducer, 34, 35, 38, 39
Poles, 81, 250, 251, 257, 259
Positive ramp, 107, 108
Power measurement, 188, 189
Power series, 198, 199
Programmable operational amplifiers, 126, 143–147
Pull-in range, 213
Pull-in time, 212, 214
Pulse position modulation, 95
Pulse width modulation, 99

Quadrature oscillator, 194–195

Ramp generator, 107
Reference terminal, 11, 13–17
Rejection bandwidth, 78
Ringing, 255, 256
Root mean square, 185–188

Scale setting, 158
Scaling factor, 139, 165, 176, 186, 199, 201
   errors in, 169, 170
Sense terminal, 11, 12, 14–17
Sequential timing, 98, 99
Settling time, 256
Signal gain, 47

Sine shaping, 110, 199–202
Single cycle generator, 105
Slewing rate, 180
Spectrum analysis, 203–206
Spurious response, 204
Square-rooting, 184, 185
Square wave generator, 94, 101
Squaring, 180, 181
Step response, 253–256
Stop band, 81

Tchebyscheff filter, 82–84
Temperature stabilisation, 122, 124, 134
Timing devices, 91–101
   dual, 99, 100
   free running, 91–95
   sequential, 98, 99
   time periods, 93–96
   triggered, 96–101
Tracking, 210, 213, 223
Transconductance, 119, 120
Transconductance amplifier, 126
Transducers, 1, 2
Transfer characteristic of PLL, 214, 218, 219
Transfer curve, 171, 218, 219
Transfer functions, 250–260
   band pass, 258
   first order high pass, 257
   first order low pass, 250
   second order high pass, 257
   second order low pass, 251
Transfer operator, 249
Transient response, 80, 253–256
Triggered operation, 96–101, 115
Triple operational amplifier, 126, 127
Twin T, 76–78
Two channel amplifier, 128, 148

Variable duty cycle square wave, 101
Variable transconductance devices, 119–141
Vector error, 171
Vector sums and differences, 194, 196–198
Voltage controlled oscillator (VCO), 101, 110, 208, 209, 215, 222, 223, 236, 239
Voltage controlled quadrature oscillator, 194, 195
Voltage controlled voltage source (VCVS), 54–64
Voltage measurements, 23, 39–42
Voltmeter
   battery powered, 41, 42
   differential, 41, 42,
   high resistance, 40, 41

Waveform generators, 101–116
   gated, 105–107, 115
   triggered, 105, 106, 115
Wide band amplifier, 128, 148, 151
Wien bridge oscillator, 154, 193

Zeros, 250, 257, 259